NATURAL DISASTER RESEARCH, PREDICTION AND MITIGATION

HURRICANES, WILDFIRES AND FLOODING

DISASTER ASSISTANCE AND CONTRACTING

Natural Disaster Research, Prediction and Mitigation

Additional books and e-books in this series can be found on Nova's website under the Series tab.

NATURAL DISASTER RESEARCH, PREDICTION AND MITIGATION

HURRICANES, WILDFIRES AND FLOODING

DISASTER ASSISTANCE AND CONTRACTING

LYDIE YOHAN
EDITOR

New York

Library of Congress Cataloging-in-Publication Data

ISBN: 978-1-53616-362-9

Published by Nova Science Publishers, Inc. † New York

CONTENTS

Preface

Chapter 1 provides a short overview of issues Congress may consider in its oversight of the Federal Emergency Management Agency's (FEMA's) federal assistance during the 2017 hurricane season (e.g., Harvey, Irma, and Maria) and other disasters (e.g., fires in California). In 2017, Hurricanes Irma and Maria damaged much of the electricity grids' transmission and distribution systems in USVI and Puerto Rico. Chapter 2 provides information on federal support for restoring the electricity grids in Puerto Rico and USVI and factors affecting this support. In 2017 two major hurricanes – Irma and Maria – caused extensive damage throughout Puerto Rico. Chapter 3 describes FEMA's Public Assistance spending in Puerto Rico and oversight efforts of federal recovery funds, and initial challenges with the recovery process. Chapter 4 provides information on DRF funding provided to Puerto Rico as a result of assistance associated with a major disaster. The primary focus of the territorial and federal efforts thus far has largely been on restoring electric power in Puerto Rico as reported in chapter 5. In September 2017, two major hurricanes—Irma and Maria—struck the USVI, causing billions of dollars in damage to its infrastructure, housing, and economy. Chapter 6 describes the status of FEMA's Public Assistance program funding provided to the USVI in response to the 2017 hurricanes as of October 1, 2018, and the USVI's transition to implementing the Public Assistance alternative procedures in the territory. Chapter 7 provides information on DRF funding provided to the U.S. Virgin Islands as a result of assistance associated with a major disaster. Chapter 8 provides information on DRF funding provided to Florida as a result of assistance associated with a major disaster. Chapter 9 provides a brief overview of the major disaster declaration process and federal assistance programs potentially available to those affected by the 2019 flooding in the Midwest. Following Hurricane Katrina, Congress required FEMA to establish advance contracts for goods and services to enable the government to quickly and effectively mobilize resources in the aftermath of a disaster. Chapter 10 assesses FEMA and USACE's use of advance contracts, FEMA's planning and reporting of selected advance contracts, and challenges, if any, with FEMA's use of these contracts. Chapter 11 addresses the extent to which federal agencies obligated funds on post-disaster contracts in response to the these events, and selected agencies experienced challenges in the planning of selected contracts.

Chapter 1 - This chapter provides a short overview of issues Congress may consider in its oversight of the Federal Emergency Management Agency's (FEMA's) federal assistance during the 2017 hurricane season (e.g., Harvey, Irma, and Maria) and other disasters (e.g., fires in California). For the current status of response efforts, see official government sources and news media.

Chapter 2 - In 2017, Hurricanes Irma and Maria damaged much of the electricity grids' transmission and distribution systems in USVI and Puerto Rico. The hurricanes left most of USVI's 106,405 people and all of Puerto Rico's 3.3 million without power and resulted in the longest blackout in U.S. history. Under the National Response Framework, electric utilities are responsible for repairing infrastructure and restoring service. They often use mutual assistance—voluntary partnerships with other electric utilities—to bring in additional resources to help restore electricity. Federal agencies provide financial assistance; help coordinate the federal response; and in severe emergencies, provide logistical support, such as assisting in damage assessments and location and transportation of repair crews and equipment. GAO was asked to review the federal response to the 2017 hurricanes. This chapter provides information on federal support for restoring the electricity grids in Puerto Rico and USVI and factors affecting this support. GAO has ongoing work examining federal support to improve grid resilience in Puerto Rico. GAO reviewed agency documents and funding data through July 20, 2018, the most recent data available; interviewed officials from FEMA, DOE, and USACE; and conducted site visits to Puerto Rico and USVI.

Chapter 3 - In 2017 two major hurricanes – Irma and Maria – caused extensive damage throughout Puerto Rico. Hurricane Maria, a Category 4 hurricane, was the most intense hurricane to make landfall in Puerto Rico since 1928, destroying roads, buildings, and cutting power and communication lines, among other things. Puerto Rico estimates that $132 billion will be needed to repair and reconstruct infrastructure and services. FEMA—a component of the Department of Homeland Security (DHS)—is the lead federal agency responsible for assisting Puerto Rico as it recovers. FEMA administers the Public Assistance program in partnership with Puerto Rico to provide funds to rebuild damaged infrastructure and restore critically-needed services. GAO was asked to review the federal government's recovery efforts related to the 2017 hurricanes. This chapter, among other objectives, describes (1) FEMA's Public Assistance spending in Puerto Rico and oversight efforts of federal recovery funds, and (2) initial challenges with the recovery process. GAO reviewed Public Assistance program documents; analyzed grant funding data; and interviewed officials from Puerto Rico and DHS about the Public Assistance program and recovery efforts, as well as officials from ten municipalities selected on the basis of population and Public Assistance spending. GAO is not making recommendations at this time, but will continue monitoring the recovery as part of its ongoing work.

Chapter 4 - The Disaster Relief Fund (DRF) is managed by the Federal Emergency Management Agency (FEMA) and is the primary source of federal assistance following a major disaster declaration. The authority to declare a major disaster is provided to the President under the Robert T. Stafford Emergency Relief and Disaster Assistance Act (P.L. 93-288, as amended; Stafford Act). The information below includes only DRF funding provided to Puerto Rico as a result of assistance associated with a major disaster declaration declared between FY2000 and FY2016. This information does not include other assistance provided through the DRF or disaster assistance provided by other federal sources, such as Small Business Administration disaster loans. In FY2017, two major disasters were declared for Puerto Rico. Those major disaster declarations are not included in the analysis below. FEMA administers disaster relief through regional offices around the country. Puerto Rico is part of Region II, headquartered in New York, NY. Puerto Rico is considered a state under Section 102 of the Stafford Act (42 U.S.C. 5122) for the purposes of that act.

Chapter 5 - On September 20, 2017, Hurricane Maria made landfall in Puerto Rico as a Category 4 storm with sustained wind speeds of over 155 miles per hour. At that time, the Commonwealth of Puerto Rico was already in recovery mode following the glancing blow struck by Hurricane Irma on September 6, 2017, which left 70% of electricity customers without power. Puerto Rico's office of emergency management reported that Hurricane Maria had incapacitated the central electric power system, leaving the entire island without power as the island's grid was essentially destroyed. Even before the 2017 hurricane season, Puerto Rico's electric power infrastructure was known to be in poor condition, due largely to underinvestment and the perceived poor maintenance practices of the Puerto Rico Electric Power Authority (PREPA). The primary focus of territorial and federal efforts thus far has largely been on restoring electric power in Puerto Rico. The new hurricane season in the Atlantic Basin (comprised of the Atlantic Ocean, the Caribbean Sea, and the Gulf of Mexico) began on June 1, 2018, and will last until November 30. A large part of the Federal Emergency Management Agency's (FEMA's) role in Puerto Rico was centered on coordinating the restoration of electric power. For this task, FEMA made available about 900 portable generators (including several generation units providing over 50 Megawatts of capacity at the partially operating Palo Seco station in San Juan), and contracted with the U.S. Army Corps of Engineers (USACE) to oversee restoration of the electric transmission system, which was damaged extensively. PREPA itself has largely been focused on the restoration of the electric distribution system, and service connections to its customers. In Puerto Rico, USACE is not only restoring emergency power but also leading initial grid repair. With the end of the grid repair part of USACE's mission assignment on May 18, 2018, almost 99% of PREPA's customers have seen their electric service restored. Section 21101 of the Bipartisan Budget Act of 2018 (P.L. 115-123) provided $28 billion in supplemental appropriations to the U.S. Department of Housing and Urban Development (HUD), under the Community Development Block Grant (CDBG) fund. Of this amount, up to $2 billion was made available (until expended) for "enhanced or improved electrical power systems." Approximately $2 billion in CDBG funds were made available to "enhance or improve" any electric power systems damaged by Hurricane Maria. To help guide the process of rebuilding Puerto Rico's grid, the U.S. Department of Energy (DOE) reports that five national laboratories have collaboratively built a model of Puerto Rico's electricity system to test how to place microgrids, determine where to place power lines underground, and test siting of renewable energy projects where they can be sheltered from damage by extreme weather events. DOE believes its modeling efforts can therefore help guide HUD and FEMA CDBG investments to improve the power grid in Puerto Rico. DOE says that it plans to complete a "resilient grid model" to prioritize investments for "transmission, distribution, new generation, energy storage, microgrids, and strategic power reserves." This would allow potential impacts on other critical infrastructure such as the petroleum, natural gas, and telecommunications sectors to be estimated. With the power restoration effort almost finished, the next focus for authorities will likely be on rebuilding Puerto Rico's electricity transmission and distribution systems, as a modern system built to U.S. industry standards is more likely to survive extreme weather events. Once the backbone of the infrastructure is in place, then the major effort of making the electricity system of Puerto Rico more resilient can follow in earnest. Building resilience would likely require improvements that go beyond even modernization and rebuilding. Resilience may require additional improvements to the system aimed at better withstanding the effects of extreme weather events.

Chapter 6 - In September 2017, two major hurricanes—Irma and Maria—struck the USVI, causing billions of dollars in damage to its infrastructure, housing, and economy. FEMA—a component of the Department of Homeland Security—is the lead federal agency responsible for assisting the USVI as it recovers from these natural disasters. Among other responsibilities, FEMA administers the Public Assistance program in partnership with the USVI territorial government, providing the USVI grant funding for response and recovery activities, including debris removal efforts, life-saving emergency protective measures, and the repair, replacement, or restoration of public infrastructure. GAO was asked to review the federal government's response and recovery efforts related to the 2017 hurricanes. This chapter describes (1) the status of FEMA's Public Assistance program funding provided to the USVI in response to the 2017 hurricanes as of October 1, 2018, and (2) the USVI's transition to implementing the Public Assistance alternative procedures in the territory. GAO reviewed program documents and data on obligations and expenditures as of October 1, 2018, and interviewed officials from FEMA and the USVI regarding the Public Assistance program specifically and disaster recovery efforts more generally. GAO also conducted site visits to the USVI islands of St. Croix, St. Thomas, and St. John. GAO is not making any recommendations in this chapter, but will continue to monitor the progress of the USVI's recovery as part of its ongoing work.

Chapter 7 - The Disaster Relief Fund (DRF) is managed by the Federal Emergency Management Agency (FEMA) and is the primary source of federal assistance following a major disaster declaration. The authority to declare a major disaster is provided to the President under the Robert T. Stafford Emergency Relief and Disaster Assistance Act (P.L. 93-288, as amended; Stafford Act). The information below includes only DRF funding provided to the U.S. Virgin Islands as a result of assistance associated with a major disaster declaration declared between FY2000 and FY2016. This information does not include other assistance provided through the DRF or disaster assistance provided by other federal sources, such as Small Business Administration disaster loans. In FY2017, two major disasters were declared for the U.S. Virgin Islands. Those major disaster declarations are not included in the analysis below. FEMA administers disaster relief through regional offices. The U.S. Virgin Islands are part of Region II, headquartered in New York, NY. The U.S. Virgin Islands are considered a state under Section 102 of the Stafford Act (42 U.S.C. 5122) for the purposes of the act.

Chapter 8 - The Disaster Relief Fund (DRF) is managed by the Federal Emergency Management Agency (FEMA) and is the primary source of federal assistance following a major disaster declaration. The authority to declare a major disaster is provided to the President under the Robert T. Stafford Emergency Relief and Disaster Assistance Act (P.L. 93-288). The information below includes only DRF funding provided to Florida as a result of assistance associated with a major disaster declaration. This information does not include other assistance provided through the DRF or disaster assistance provided by other federal sources, such as Small Business Administration disaster loans. FEMA administers disaster relief through regional offices around the country. Florida is part of Region IV, headquartered in Atlanta, GA.

Chapter 9 - This Insight provides a brief overview of the major disaster declaration process and federal assistance programs potentially available to those affected by the current flooding in the Midwest. As authorized under the Robert T. Stafford Disaster Relief and Emergency Assistance Act (P.L. 93-288, as amended; 42 U.S.C. §§5121 et seq.), the President may issue a major disaster declaration due to catastrophic flooding. This allows for a broad range of federal assistance programs to be made available to state and local governments, private nonprofit organizations, and individuals through the Federal Emergency Management Agency (FEMA) and other federal agencies. According to the National Oceanic and Atmospheric Administration (NOAA), the record flooding in Nebraska, Iowa, and Minnesota is the result of "rapid snow melt combined with heavy spring rain and late season snowfall in areas where soil moisture is high." On March 21, 2019, President Donald J. Trump issued a major disaster declaration for the State of Nebraska for the severe winter storm, straight-line winds, and flooding that began on March 9, 2019. On March 23, 2019, President Trump issued a major disaster declaration for the State of Iowa for the severe storms and flooding that began on March 12, 2019. Other states affected by flooding have declared states of emergency, including Missouri, Kansas, Wisconsin, Ohio, and South Dakota as well as tribal nations in Nebraska, Iowa, and South Dakota. Additionally, February rainfall and severe storms that resulted in flooding in Mississippi and Tennessee resulted in their governors' submission of requests for major disaster declarations on March 21 and March 22, 2019, respectively. NOAA's spring outlook states that additional rain and melting snow "will prolong and expand flooding," and that as water flows downstream, the threat of flooding will increase and become more widespread.

Chapter 10 - Following Hurricane Katrina, Congress required FEMA to establish advance contracts for goods and services to enable the government to quickly and effectively mobilize resources in the aftermath of a disaster, like those that affected the United States in 2017. GAO was asked to review the federal government's response to the three 2017 hurricanes and California wildfires. This chapter assesses, among other things, (1) FEMA and USACE's use of advance contracts, (2) FEMA's planning and reporting of selected advance contracts, and (3) challenges, if any, with FEMA's use of these contracts. GAO analyzed data from the Federal Procurement Data System-Next Generation through May 31, 2018; selected a non-generalizable sample of 14 FEMA and USACE advance contracts that were competed and obligated over $50 million, or non-competed and obligated over $10 million, in response to the 2017 disasters; and interviewed FEMA and USACE officials.

Chapter 11 - Federal contracts play a key role in timely response and recovery efforts following disasters. While federal agencies, such as FEMA and USACE, may have advance contracts in place for obtaining goods and services following disasters, agencies may also award post-disaster contracts. GAO was asked to review the federal government's response to three major hurricanes in 2017, as well as the 2017 California wildfires. This chapter addresses, among other objectives, the extent to which (1) federal agencies obligated funds on post-disaster contracts in response to the these events, and (2) selected agencies experienced challenges in the planning of selected contracts. GAO analyzed data from the Federal Procurement Data System-Next Generation; selected a non-generalizable sample of 23 post-disaster contracts based on factors such as if the contract was set aside for award to a local contractor; reviewed federal regulations and agency guidance; and interviewed agency officials.

In: Hurricanes, Wildfires and Flooding
Editor: Lydie Yohan
ISBN: 978-1-53616-362-9

Chapter 1

NATURAL DISASTERS OF 2017: CONGRESSIONAL CONSIDERATIONS RELATED TO FEMA ASSISTANCE*

Jared T. Brown, Diane P. Horn, William L. Painter and Shawn Reese

ABSTRACT

This chapter provides a short overview of issues Congress may consider in its oversight of the Federal Emergency Management Agency's (FEMA's) federal assistance during the 2017 hurricane season (e.g., Harvey, Irma, and Maria) and other disasters (e.g., fires in California). For the current status of response efforts, see official government sources and news media.

* This is an edited, reformatted and augmented version of Congressional Research Service, Publication No. IN10810, dated December 6, 2017.

STAFFORD ACT DECLARATIONS AND RESPONSE

Under the Robert T. Stafford Disaster Relief and Emergency Assistance Act (the Stafford Act), the President may declare an emergency or major disaster to authorize federal assistance, if the capacities of state and tribal governments are overwhelmed. As authorized by numerous sections of the Stafford Act (e.g., §302) and the Homeland Security Act of 2002 (e.g., §504), FEMA is responsible for coordinating the federal disaster response, as guided by the National Response Framework and subcomponent policies. Many deployable federal assets have been used in responding to the 2017 disasters. Other federal agencies are frequently incorporated by, and compensated for, their federal response through Mission Assignments. Congress may evaluate whether the federal response to the 2017 disasters has been effectively led by FEMA and supported by other federal agencies. Congress can use after-action reports, like those after Hurricane Sandy, to inform oversight and reforms.

FEDERAL FINANCIAL ASSISTANCE

FEMA has multiple disaster assistance programs, including:

- The Individual Assistance (IA) Program comprised of: (1) Mass Care and Emergency Assistance, (2) Crisis Counseling Assistance and Training Program, (3) Disaster Unemployment Assistance, (4) Disaster Legal Services, (5) Disaster Case Management, and (6) the Individuals and Households Program. IA can include some or all of these programs, depending on what is requested and approved by FEMA. Congress may evaluate the numerous factors FEMA considers when it decides whether to recommend providing IA.
- The Public Assistance (PA) Grant Program provides grants to tribal, state, local governments, and certain private nonprofit

organizations to fund emergency protective services, conduct debris removal operations, and repair or replace damaged public facilities. Congress may consider the cost-share required for PA, how it complements other federal assistance, and whether FEMA's recent minimum standards policy provides sufficient mitigation against future disasters on PA-funded replacement projects.

- Hazard Mitigation Assistance (HMA) Programs encompass three separate programs: the Hazard Mitigation Grant Program (HMGP); the Pre-Disaster Mitigation program; and the Flood Mitigation Assistance program. HMGP receives significantly more funding following major disaster declarations than the other two programs. Congress may consider if these HMA programs are sufficiently coordinated with other federal programs and resourced to adequately reduce future disaster risk.
- Fire Mitigation Assistance Grants (FMAGs) provide various forms of federal fire suppression assistance for "declared" fires that might become a major disaster. Congress may consider whether this assistance, coupled with other federal fire assistance, is sufficient to address a perceived increase in wildfire risk, and whether federal wildfire funding mechanisms should be reformed.

These FEMA grant programs all have state cost-share requirements. PA cost-shares can be adjusted by the Administration, but IA and HMGP cost-shares are set in law. Congress has adjusted cost-shares for specific states and disasters, such as after the 2005 Gulf Coast hurricanes. Recent major disaster declarations have had various cost-share adjustments for PA. Congress may consider legislation to revise these programs, such as H.R. 4460, which was ordered reported from the House Transportation and Infrastructure Committee.

FEMA also administers non-grant financial assistance:

- The National Flood Insurance Program (NFIP) is the primary source of flood insurance coverage for residential properties. NFIP has implemented temporary changes to the claims process to allow policyholders to receive funds more quickly in some of the areas affected by 2017 floods. However, past data on participation rates suggest that many properties in the Special Flood Hazard Areas (SFHAs) affected by the hurricanes may not have flood insurance. By law, federal assistance to owners of these uninsured properties is more restricted than to owners of properties with insurance or those living outside the SFHA.

- Key provisions of the NFIP were reauthorized through December 8, 2017 (P.L. 115-56, Division D, §130). OMB has requested revisions to the NFIP. Reform ideas may be considered by Congress before these authorities expire.
- The Community Disaster Loan (CDL) Program provides loan assistance to governments to compensate for post-disaster loss of revenues. Recent appropriations provisos for CDLs afford the Administration discretion to provide CDLs, so past guidance on CDLs may not apply to future loans.

FUNDING

FEMA's Disaster Relief Fund (DRF) is the primary funding source for immediate FEMA response and relief in the wake of disasters. As Hurricane Harvey approached on August 25, 2017, the DRF had approximately $3.5 billion on hand. On August 28, FEMA implemented an "immediate needs funding restriction," limiting obligations from the DRF for some longer-term recovery and mitigation projects to preserve DRF balances for response needs. P.L. 115-56 provided $7.4 billion in supplemental appropriations for the DRF, and the continuing resolution in the act provided additional resources, so the funding restriction was lifted on October 2. $18.67 billion more was appropriated in P.L. 115-72, although a $4.9 billion transfer to the Disaster Assistance Direct Loan Program account (which funds CDLs) and a $10 million transfer to the DHS Office of Inspector General reduced the effective appropriation to $13.76 billion. The Administration submitted a third supplemental appropriations request on November 17, 2017, for $44 billion, including $23.5 billion for the DRF.

The NFIP was not designed to retain funding to cover claims for extreme events; instead, the law allows it to borrow money from the Treasury for such events. FEMA borrowed $5.285 billion from the Treasury to meet initial claims from Hurricane Harvey, and thus reached the borrowing limit of $30.425 billion. As the Administration requested, P.L. 115-72 cancelled $16 billion of NFIP debt, reducing it to $14.425 billion. FEMA borrowed another $6.1 billion on November 9, bringing the debt to $20.525 billion. The NFIP currently has $9.9 billion of remaining borrowing authority.

In: Hurricanes, Wildfires and Flooding ISBN: 978-1-53616-362-9
Editor: Lydie Yohan

Chapter 2

2017 Hurricane Season: Federal Support for Electricity Grid Restoration in the U.S. Virgin Islands and Puerto Rico*

United States Government Accountability Office

Abbreviations

DHS	Department of Homeland Security
DOD	Department of Defense
DOE	Department of Energy
FEMA	Federal Emergency Management Agency
FPDS-NG	Federal Procurement Data System-Next Generation
PREPA	Puerto Rico Electric Power Authority
PROMESA	Puerto Rico Oversight, Management, and Economic Stability Act

---

* This is an edited, reformatted and augmented version of the United States Government Accountability Office Report to Congressional Requesters, Publication No. GAO-19-296, dated April 2019.

USACE U.S. Army Corps of Engineers
VIWAPA Virgin Islands Water and Power Authority

Why GAO Did This Study

In 2017, Hurricanes Irma and Maria damaged much of the electricity grids' transmission and distribution systems in USVI and Puerto Rico. The hurricanes left most of USVI's 106,405 people and all of Puerto Rico's 3.3 million without power and resulted in the longest blackout in U.S. history.

Under the National Response Framework, electric utilities are responsible for repairing infrastructure and restoring service. They often use mutual assistance—voluntary partnerships with other electric utilities—to bring in additional resources to help restore electricity. Federal agencies provide financial assistance; help coordinate the federal response; and in severe emergencies, provide logistical support, such as assisting in damage assessments and location and transportation of repair crews and equipment.

GAO was asked to review the federal response to the 2017 hurricanes. This chapter provides information on federal support for restoring the electricity grids in Puerto Rico and USVI and factors affecting this support. GAO has ongoing work examining federal support to improve grid resilience in Puerto Rico.

GAO reviewed agency documents and funding data through July 20, 2018, the most recent data available; interviewed officials from FEMA, DOE, and USACE; and conducted site visits to Puerto Rico and USVI.

What GAO Found

Federal agencies supported efforts to restore electricity in the U.S. Virgin Islands (USVI) and Puerto Rico through the types of support they traditionally provide following disasters and, in Puerto Rico, in some unprecedented ways.

- *USVI.* Federal agencies provided traditional federal support to the electric utility's restoration efforts. For example, the Federal Emergency Management Agency (FEMA) provided financial assistance through its Public Assistance Program, and the Department of Energy (DOE) provided subject matter expertise to assist the local utility. In addition, the U.S. Army Corps of Engineers (USACE) provided generators for hospitals and other critical facilities. FEMA obligated about $795 million for these efforts as of July 20, 2018. According to the local utility, it took about 5 months for power to be restored to all customers with structures deemed safe for power restoration.
- *Puerto Rico.* In addition to the traditional types of support, FEMA and USACE undertook unprecedented roles of helping to coordinate and directly assist with grid restoration in Puerto Rico. FEMA requested that USACE lead federal grid repair efforts because of the scale of the damage and because the Puerto Rico Electric Power Authority (PREPA) did not have the capacity to respond, according to FEMA officials. FEMA obligated about $3.2 billion for electricity restoration efforts as of July 20, 2018, and PREPA estimated that it took roughly 11 months for power to be restored to all customers with structure structures deemed safe for power restoration.

Various factors affected federal support for electricity grid restoration, according to officials GAO interviewed and documents reviewed. For example, getting the crews and materials needed to islands was more difficult and time-consuming than on the mainland. In Puerto Rico, PREPA was insolvent, which presented challenges for restoring the grid. For example, PREPA canceled its vegetation management program; this contributed to the destruction of the grid when the hurricane arrived, according to FEMA officials. In addition, FEMA did not anticipate or plan for the extensive federal role in grid restoration in Puerto Rico, and USACE did not have a contract in place to immediately initiate grid repair efforts, according to USACE officials. FEMA and USACE identified potential actions to address these challenges, such as reviewing advance contracts.

April 18, 2019
Congressional Requesters

In 2017, Hurricanes Irma and Maria damaged much of the electricity grids' transmission and distribution systems in Puerto Rico and the U.S. Virgin Islands, leaving millions without power and resulting in the longest blackout in U.S. history. These hurricanes brought some of the heaviest rain and strongest winds ever recorded to Puerto Rico and the U.S. Virgin Islands, causing unprecedented damage and the costliest hurricane season on record. Following the hurricanes, it took roughly 5 months for power to be restored to all of the customers in the U.S. Virgin Islands, which has an estimated population of 106,405 people, and roughly 11 months for power to be restored to all of the customers in Puerto Rico, which has an estimated population of 3.3 million people. According to the Virgin Islands Water and Power Authority (VIWAPA) and the Puerto Rico Electric Power Authority (PREPA), power has been restored to all customers with structures deemed safe for power restoration. This does not mean that all pre-storm customers have power, as some structures may not have been deemed safe for power restoration.

Under the federal government's National Response Framework, which describes how the federal government, states and localities, and other public and private sector institutions should respond to disasters and emergencies, electric utilities are responsible for repairing damaged electricity infrastructure and restoring services. Electric utilities often use mutual assistance—voluntary partnerships with other electric utilities—to bring in additional resources beyond those of the affected utility to help restore electricity[1]. Under the National Response Framework, when appropriate, federal agencies provide financial assistance for response and recovery activities, help coordinate the response, gather and share information, and communicate with key stakeholders and the public. Also, in severe emergencies, federal agencies can provide some logistical support, such as assisting in damage assessments and locating and transporting repair crews and equipment.

The National Response Framework designates the Federal Emergency Management Agency (FEMA), a component of the Department of Homeland Security, to lead the coordination of federal disaster response efforts. The framework's 14 emergency support functions are the federal government's primary coordinating structure for building, sustaining, and delivering response capabilities during an emergency. For each of the 14 emergency support functions, a federal department or agency serves as the designated coordinator. The framework designates the Department of Energy (DOE) as the coordinating agency for federal efforts in the energy sector, which includes electricity.[2] In this role, DOE is responsible for coordinating with FEMA and other relevant federal agencies, such as the U.S. Army Corps of Engineers (USACE), and for collaborating with critical infrastructure owners and operators to prioritize and coordinate federal efforts. USACE is the coordinating agency for the public works and engineering emergency support function under the framework, which

---

[1] According to the Edison Electric Institute, the electric utility industry relies on mutual assistance networks to help speed restoration following major outages. Mutual assistance agreements are an essential part of the electric power industry's service restoration process and contingency planning, providing participating electric utilities with resources, such as equipment and lineworkers and other electric utility personnel, to help restore power when assistance is needed. These agreements provide a formal process for electric utilities to request support from other electric companies in parts of the country that have not been affected by major outages. Electric utilities affected by a major outage are thereby able to increase the size of their workforce by borrowing restoration workers from other companies. When called upon, a company will send skilled restoration workers—both company employees and contractors—along with specialized equipment to help with the restoration efforts of a fellow company.

[2] DOE is the coordinating agency for emergency support function 12, which is the energy-focused emergency support function under the National Response Framework.

includes providing support for temporary emergency power, among other things.[3]

You asked us to review the federal government's response to the 2017 hurricanes. Our objective in this chapter was to describe federal support provided to help restore the electricity grids in Puerto Rico and the U.S. Virgin Islands in response to the 2017 hurricane season and the factors that affected this support. This is our first report related to electricity infrastructure in response to your request; we have ongoing work examining federal support to improve grid resilience in Puerto Rico.

To conduct this work, we reviewed documentation related to federal agency support to help restore electricity grids in response to the 2017 hurricane season in Puerto Rico and the U.S. Virgin Islands. Specifically, we reviewed the National Response Framework; FEMA mission assignment documents; and agency reports identifying lessons learned, such as FEMA's 2017 Hurricane Season FEMA After-Action Report, DOE's after action report for the 2017 hurricane season, and USACE's 2018 Remedial Action Program Senior Leader Briefing. We also reviewed documents related to FEMA's Public Assistance Program including FEMA's Public Assistance Program and Policy Guide and FEMA's Public Assistance Alternative Procedures (Section 428) Guide for Permanent Work in Puerto Rico.

We reviewed FEMA data on obligations for electricity restoration efforts in Puerto Rico and the U.S. Virgin Islands as of July 20, 2018, the most recent data available at the time of our review.[4] FEMA officials compiled obligations related to electricity restoration from two databases, the Emergency Management Mission Integrated Environment and the Enterprise Coordination and Approvals Processing System. To assess the

---

[3] USACE is the coordinating agency for emergency support function 3, public works and engineering, which includes the provision of assets and services related to public works and engineering, such as temporary power. USACE assists FEMA by providing temporary emergency power, including the procurement, installation, operation and maintenance of generators, to critical public facilities when commercial power is interrupted by disasters.

[4] An obligation is a definite commitment that creates a legal liability of the government for the payment of goods and services ordered or received or a legal duty on the part of the United States that could mature into a legal liability by virtue of actions on the part of the other party beyond the control of the United States.

reliability of these data, we reviewed existing information about the data and discussed data quality control procedures with FEMA officials. We determined that the data we used from these systems were sufficiently reliable for the purposes of this chapter. To identify USACE contract obligations for grid restoration in Puerto Rico, we reviewed Federal Procurement Data System-Next Generation (FPDS-NG) data through June 30, 2018, the most recent data available at the time of our review.[5] We assessed the reliability of FPDS-NG data by reviewing existing information about the FPDS-NG system and the data it collects, and compared FPDS-NG data to the contract files in our review. We determined that the FPDS-NG data were sufficiently reliable for the purposes of this chapter. We also reviewed PREPA's data on the number of lineworkers working on grid restoration efforts in Puerto Rico. We assessed the reliability of PREPA's lineworker data by interviewing PREPA officials on how the data were collected and maintained, and we also discussed the reliability of the data with FEMA, USACE, and DOE officials. We found these data to be sufficiently reliable for our purposes.

We interviewed federal officials from FEMA, DOE, and USACE and representatives from two relevant industry organizations—the Edison Electric Institute and the American Public Power Association.[6] We also conducted site visits to Puerto Rico and the U.S. Virgin Islands to interview representatives of local government agencies; the electric utilities, PREPA and VIWAPA; and utility commissions.

To identify any factors that affected federal support for the restoration of electricity grids in Puerto Rico and the U.S. Virgin islands, we summarized officials' and representatives' views on any factors that affected federal support for the restoration of the electricity grids. We

---

[5] For the purposes of this chapter, contract obligations include obligations against what the FPDS-NG categorizes as definitive vehicles (definitive contracts and purchase orders that have a defined scope of work that do not allow for individual orders under them) and against what FPDS-NG categorizes as indefinite delivery vehicles (including orders under indefinite delivery vehicles, such as indefinite delivery, indefinite quantity contracts).

[6] The Edison Electric Institute is an association that represents all U.S. investor-owned electric companies. The American Public Power Association is an association of not-forprofit, community-owned electric utilities.

included those factors that were most often cited in interviews and documents we reviewed.

We conducted this performance audit from February 2018 to April 2019 in accordance with generally accepted government auditing standards. Those standards require that we plan and perform the audit to obtain sufficient, appropriate evidence to provide a reasonable basis for our findings and conclusions based on our audit objectives. We believe that the evidence obtained provides a reasonable basis for our findings and conclusions based on our audit objectives.

BACKGROUND

This section provides information on the electricity provider, impact of 2017 hurricanes, and status of electricity restoration in Puerto Rico and the U.S. Virgin Islands. Also, it describes FEMA's Public Assistance Program.

Puerto Rico

Electricity provider. PREPA is a public power utility owned by the Commonwealth of Puerto Rico and a monopoly supplier of electricity in the commonwealth. It is also one of the nation's largest public power utilities, serving approximately 1.5 million customers. PREPA was approximately $9 billion in debt prior to Hurricanes Irma and Maria, and its electric power infrastructure was known to be in poor condition, largely due to underinvestment and poor maintenance practices.[7] In May 2018, we

---

[7] Congressional Research Service, *Repair or Rebuild: Options for Electric Power in Puerto Rico* (Nov. 16, 2017). In July 2017, the Financial Oversight and Management Board for Puerto Rico put PREPA into a bankruptcy-like process. In June 2016, Congress enacted and the President signed the Puerto Rico Oversight, Management, and Economic Stability Act (PROMESA) Pub. L. No. 114-187, in response to Puerto Rico's fiscal crisis. PROMESA established the Financial Oversight and Management Board for Puerto Rico, and granted it broad powers of fiscal and budgetary control over Puerto Rico. PROMESA also established a mechanism through which the Oversight Board could petition U.S. courts on Puerto Rico's behalf to restructure debt. Under federal bankruptcy laws, Puerto Rico is otherwise ineligible to authorize its municipalities and instrumentalities to petition U.S. courts to

found that inadequate management of PREPA's financial condition contributed to Puerto Rico's persistent deficits.[8] Specifically, PREPA did not update or improve its electric generation and transmission systems, which hampered their performance and led to increased costs of producing electricity that it did not fully pass on to consumers. In addition, Puerto Rico's economy is in a prolonged period of economic contraction, and according to U.S. Census Bureau estimates, its population declined from a high of approximately 3.8 million people in 2004 to 3.3 million people in 2017, a decline of 12.8 percent. Along with the declining population, demand for electricity declined 18 percent from 2007 to 2017, according to PREPA.[9]

Impact of 2017 hurricanes. Hurricanes Irma and Maria in September 2017 left Puerto Rico's entire electricity grid inoperable, according to the economic and disaster recovery plan for Puerto Rico.[10] According to a report by a working group that included utilities and national laboratories, among others, because of the extended and unprecedented damage, a significant portion of the generation, transmission, and distribution system must be rebuilt, including high-voltage transmission lines that often survive lower category hurricanes.[11] While Puerto Rico's population had already been declining, the migration of people from Puerto Rico accelerated following the hurricanes, according to PREPA.

---

restructure debt. For more information on PROMESA, see GAO *Puerto Rico: Factors Contributing to the Debt Crisis and Potential Federal Actions to Address Them*, GAO-18-387 (Washington, D.C.: May 9, 2018).

[8] GAO-18-387.

[9] Puerto Rico Electric Power Authority, *Puerto Rico Electric Power Authority Fiscal Plan* (Aug. 1, 2018).

[10] Central Office for Recovery, Reconstruction, and Resiliency, Transformation and Innovation in the Wake of Devastation: An Economic and Disaster Recovery Plan for Puerto Rico (August 8, 2018.

[11] Puerto Rico Energy Resiliency Working Group and Navigant Consulting, Inc., *Build Back Better: Reimagining and Strengthening the Power Grid of Puerto Rico* (Dec. 11, 2017). The working group included the New York Power Authority; Puerto Rico Electric Power Authority; Puerto Rico Energy Commission; Consolidated Edison Company of New York; Inc.; Edison International; Electric Power Research Institute; Long Island Power Authority; Smart Electric Power Alliance; DOE; Brookhaven National Laboratory; National Renewable Energy Laboratory; Pacific Northwest National Laboratory; Grid Modernization Lab Consortium; and Public Service Electric and Gas Long Island.

Status of electricity restoration. According to PREPA, it took roughly 11 months for power to be restored to all of the customers able to receive power safely in Puerto Rico following the hurricanes. A PREPA official told us that PREPA's estimates of customers with power restored are based on the number of meters that it knows are served by a given power line and on the number of meters it can read currently. Power has been restored to 100 percent of customers that are able to receive power safely, but this does not mean that all pre-storm customers have power restored, as some structures may not have been deemed safe for power restoration, according to PREPA officials. Figure 1 shows the percentage of customers with electricity restored in Puerto Rico beginning in January 2018 when PREPA was able to start estimating this information.

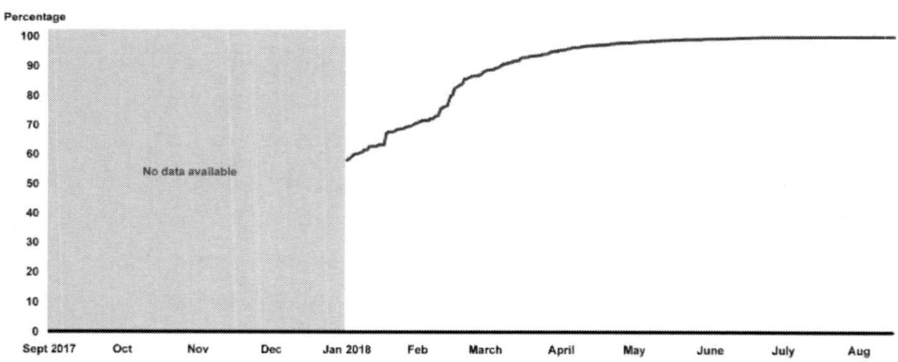

Source: Puerto Rico Electric Power Authority (PREPA) estimates | GAO-19-296.

Note: PREPA's estimates of customers with power restored are based on the number of meters that PREPA knows are served by a given power line and on the number of meters PREPA can read currently, according to a PREPA official. Power has been restored to 100 percent of customers that are able to receive power safely, but this does not mean that all pre-storm customers have power restored, as some structures may not have been deemed safe for power restoration, according to PREPA officials.

Figure 1. Estimated Percentage of Customers Able to Receive Electricity with Electricity Restored in Puerto Rico from January 2018 through August 2018.

Although PREPA estimates that electricity had been restored to all customers by August 2018, in some instances electricity service has been supported by temporary generators, and outages have continued. For example, as of December 11, 2018, USACE was supporting seven generators that were supporting micro grids for the island municipalities of Vieques and Culebra. These islands had previously been served by an undersea transmission line connecting the islands to PREPA's main grid on Puerto Rico. According to the U.S. Energy Information Administration, total electricity sales in Puerto Rico returned to pre–Hurricane Maria levels as of April and May 2018, although residential electricity sales appear to continue to lag historical levels, reflecting some continued outages.

U.S. Virgin Islands

Electricity provider. VIWAPA, a public utility, is a monopoly provider of electric power services in the U.S. Virgin Islands and serves approximately 55,000 customers throughout the territory. Like PREPA, VIWAPA faced financial challenges before the hurricanes. The USVI Hurricane Recovery and Resilience Task Force Report noted that VIWAPA has a 17 percent non-payment rate across its customer base, a significant unfunded pension liability, and long-term debt commitments of $265 million.[12] In addition, the report states that the U.S. Virgin Island's energy system faces many challenges that have led to higher rates and a historically unreliable grid. These include an aging, inefficient, and oversized infrastructure and heavy reliance on imported fossil fuels. The report also says that peak demand declined 18 percent from 2011 through 2017, driven by a variety of factors, including population decline. In addition, the report says that VIWAPA's high energy rates and reliability issues have led some customers— particularly larger commercial and industrial ones—to leave the grid.

Impact of 2017 hurricanes. Hurricanes Irma and Maria damaged more than 90 percent of VIWAPA's aboveground power lines and over 20 percent of VIWAPA's generation capacity, according to the USVI Hurricane Recovery and Resilience Task Force Report. Specifically, the

---

[12] USVI Hurricane Recovery and Resilience Task Force, *USVI Hurricane Recovery and Resilience Task Force Report* (2018).

hurricanes damaged more than 20,000 poles and 1,100 miles of transmission and distribution lines, according to the report. Although 90 percent of VIWAPA's above ground power lines were damaged, this was far fewer than the miles of transmission and distribution lines damaged in Puerto Rico.

Electricity status. According to VIWAPA, following the hurricanes, it took roughly 5 months for power to be restored to all of the eligible customers in the U.S. Virgin Islands. Eligible customers were those whose homes were safely able to receive power. Some homes had suffered substantial damage to their electrical infrastructure from the hurricanes and were not able to receive power safely until their electrical equipment was repaired. VIWAPA's estimates of customers with power restored are based on the number of meters that VIWAPA knows are served by a given power line, as VIWAPA's automated system for determining the percentage of customers without power was destroyed and is still being restored, according to a FEMA official. Although electricity service has been restored, electricity demand has not recovered to prestorm levels. According to the USVI Hurricane Recovery and Resilience Task Force Report, VIWAPA's peak demand—the maximum energy load consumed by customers at any point in a year—was approximately 107 megawatts before the storms, but as of May 2018 it was 66 megawatts. The report says that demand will likely rebound to some degree as the territory rebuilds and recovers; however, it is unclear how quickly or by how much.

FEMA' s Public Assistance Program

F assistance through its Public Assistance Program to state, EMA, in leading the coordination of federal disaster response efforts, provides territorial, local, and tribal governments and certain types of private nonprofit organizations to assist them in responding to and recovering from major disasters or emergencies.

FEMA Public Assistance Program funds can be provided for

- emergency work, such as for emergency protective measures that must be done immediately to protect public health and safety;
- permanent work, which includes the restoration of disaster-damaged public utilities;[13] and
- management costs, which include indirect costs, administrative expenses, or other expenses that are not directly chargeable to a specific project and that a recipient or subrecipient incurs in administering and managing Public Assistance awards.

Generally, emergency work takes place for about 6 months following a disaster, while permanent work can take place over a decade, according to FEMA officials. FEMA can provide grants for both emergency and permanent work, and it can also provide direct federal assistance for emergency work. Under direct federal assistance, federal agencies directly perform or contract for the emergency work. FEMA's Public Assistance Program allows for the federal government to provide direct assistance at the request of the state, territorial, and local governments when the impact of an incident is so severe that the state, territorial, and local governments lack the capability to perform or contract eligible emergency work. Under the Public Assistance Program and the Stafford Act, FEMA may mission assign—issue a work order that directs another federal agency, such as DOE or USACE to utilize its authorities and the resources granted to it under federal law—in support of this direct assistance to state, local, and territorial governments.[14]

---

[13] Grants for permanent work may also include cost-effective hazard mitigation to protect the facilities from future damage.

[14] See 42 U.S.C. § 5192(a)(1); 44 C.F.R. § 206.2(a)(18). Mission assignments may be categorized as direct federal assistance when FEMA tasks agencies to provide eligible emergency work or debris removal services to state, local, tribal, or territorial governments following an emergency or major disaster declaration.

FEMA's Community Disaster Loan Program

The Community Disaster Loan program provides loans to local governments that have suffered substantial loss of tax and other revenue in areas included in a major disaster declaration. The loan funding may be used for existing essential municipal functions and expanded functions required to meet disaster-related needs, but not for capital improvements or repair or restoration of damaged public facilities.

THE FEDERAL ROLE IN ELECTRICITY GRID RESTORATION WAS UNPRECEDENTED IN PUERTO RICO, AND VARIOUS FACTORS AFFECTED THE SUPPORT PROVIDED IN PUERTO RICO AND THE U.S. VIRGIN ISLANDS

Federal agencies provided traditional support to restore electricity in response to Hurricanes Irma and Maria in both Puerto Rico and the U.S. Virgin Islands—such as providing temporary power for critical facilities. They also provided unprecedented support in Puerto Rico by helping to coordinate efforts to repair Puerto Rico's electricity grid rather than primarily supporting the local utility's efforts. Factors that affected the electricity grid restoration efforts in Puerto Rico and the U.S. Virgin Islands included logistical constraints, availability of materials, the financial condition of local utilities, and the unprecedented and extensive role of federal agencies. Appendix I provides timelines of federal and other efforts to support electricity grid restoration in Puerto Rico and the U.S. Virgin Islands after the 2017 hurricane season.

Federal Support Provided to Restore Electricity in Puerto Rico and the U.S. Virgin Islands in Response to the 2017 Hurricanes Included an Unprecedented Role for the Federal Government

Federal agencies assisted in the restoration of electricity after Hurricanes Irma and Maria in a variety of ways. FEMA provided billions in grants and direct federal assistance for electricity restoration. DOE provided subject matter expertise and coordination assistance. USACE provided temporary emergency power in Puerto Rico and the U.S. Virgin Islands. In addition, FEMA and USACE undertook unprecedented roles to help coordinate and directly assist with grid restoration in Puerto Rico.

Grants, Direct Federal Assistance, and Loans from FEMA

FEMA provided billions in grants and direct federal assistance to support electricity restoration in Puerto Rico and the U.S. Virgin Island through its Public Assistance Program. As public utilities, both PREPA and VIWAPA are eligible applicants for federal assistance through FEMA's Public Assistance Program for the repair, restoration, and replacement of public facilities damaged or destroyed by a major disaster.[15] As of July 20, 2018, FEMA had obligated approximately $3.2 billion for direct federal assistance through mission assignments and Public Assistance grant funds for electricity restoration in Puerto Rico and approximately $795 million for the U.S. Virgin Islands.[16] This includes $2

[15] Generally, under the Public Assistance Program, an eligible applicant who owns a public facility or facilities may receive grant assistance to repair or replace the facility to its predisaster design (its size and capacity) and function. Grants additionally cover eligible expenses that are required by current applicable codes and standards, provided the work is required as a direct result of the disaster. In addition, the Bipartisan Budget Act of 2018 includes a provision that for the duration of the recovery from Hurricanes Irma and Maria in Puerto Rico and the U.S. Virgin Islands, FEMA may provide, for critical services such as power, as defined in the Stafford Act, assistance to replace or restore the function of a facility or system to industry standards without regard to the predisaster condition of the facility.

[16] This includes funding for Public Assistance grants and direct federal assistance mission assignments that FEMA officials identified as related to power restoration, including temporary emergency power. It does not include funding for federal operational support mission assignments, which are those that task federal agencies with providing federal-to-federal support.

billion that FEMA obligated for direct federal assistance through mission assignments to USACE for temporary emergency power and grid restoration efforts in Puerto Rico. In the U.S. Virgin Islands, FEMA obligated $63 million for direct federal assistance related to electricity restoration, most of which was obligated to USACE and DOE. Table 1 shows FEMA funding obligations for electricity restoration efforts in Puerto Rico and the U.S. Virgin Islands.

Table 1. Federal Emergency Management Agency Public Assistance obligations for electricity restoration in puerto rico and The U.S. Virgin Islands following hurricanes Irma and Maria, as of July 20, 2018

	Grants (dollars)	Direct federal assistance (dollars)[a]	Total (dollars)[b]
Puerto Rico	1,195,861,552	2,003,600,000	3,199,461,552
U.S. Virgin Islands	731,943,655	63,070,000	795,013,655

Source: GAO analysis of Federal Emergency Management Agency (FEMA) data. | GAO-19-296.

[a] Direct federal assistance is provided when FEMA issues a mission assignment for federal agencies to provide support when states or territories lack the resources to undertake emergency work and they request such assistance.

[b] In addition to federally provided funds, states and territories may have to pay some portion of the total cost, or the President may waive the cost share. In Puerto Rico the President authorized 100 percent federal funding for grid restoration until August 20, 2018, and for emergency temporary power support until September 19, 2018. In the U.S. Virgin Islands, the President authorized 100 percent federal funding until May 14, 2018, for emergency protective measures and September 15, 2018, for emergency temporary power support. In addition, the Insular Areas Act provides an administering agency discretion to waive any requirement for matching funds to be provided by the U.S. Virgin Islands or other specified territories.

In addition, FEMA provided $75 million to VIWAPA through the Community Disaster Loan program as of July 20, 2018, according to FEMA officials. FEMA officials said that the most common use for Community Disaster Loan funds is payroll, and other examples of eligible uses include employee benefits, facilities maintenance costs, and normal operating materials.

Coordination and Technical Assistance from DOE

DOE received mission assignments from FEMA and deployed staff from its headquarters, site offices, and power marketing administrations to provide subject matter expertise and technical assistance in support of electricity grid damage assessments and power restoration efforts in both Puerto Rico and the U.S. Virgin Islands. According to DOE officials, DOE's primary role in the response efforts on Puerto Rico and the U.S. Virgin Islands was coordination and provision of subject matter experts, as is typical for DOE's role as the lead agency for the energy sector emergency support function. In Puerto Rico, however, DOE provided more advisors for a longer period of time than would be typical because of the extent of the damage to the electricity grid in Puerto Rico and PREPA's limited capacity to respond, according to DOE officials. Specifically, DOE staffed up to 12 project support advisors to Puerto Rico from October 18, 2017, to August 8, 2018, and one supply chain support advisor from December 18, 2017, to March 16, 2018. These advisors provided subject matter expertise to USACE by reviewing construction plans and providing recommendations for prioritization, and scheduling and assisting in inventory management for incoming electrical grid equipment, among other things, according to DOE. In addition, in the U.S. Virgin Islands DOE deployed a team of 36 people from the Western Area Power Administration along with trucks and materials to help rebuild the electricity grid through a FEMA mission assignment.[17] DOE officials told us that the department is also providing ongoing support on how to improve grid resilience as part of grid restoration and recovery efforts in both Puerto Rico and the U.S. Virgin Islands.[18]

---

[17] The Western Area Power Administration is one of four power marketing administrations within DOE that operate electric systems and sell electricity generated at federally owned and operated hydroelectric dams. The assistance provided by the Western Area Power Administration in the U.S. Virgin Islands was similar to that provided by mutual assistance crews, though the power administration was paid for this work through a FEMA mission assignment to DOE rather than through reimbursement from the local utility.

[18] We have ongoing work examining agencies' efforts to support grid resilience and restoration in Puerto Rico, which we will report on in 2019.

Temporary Power from USACE

USACE provided temporary emergency power for critical facilities in Puerto Rico and the U.S. Virgin Islands. These temporary emergency power missions provided and maintained generators to deliver electricity to critical public facilities, such as hospitals and relief centers.[19] After receiving a FEMA mission assignment to provide temporary emergency power in Puerto Rico, USACE deployed its Emergency Power Planning and Response Team, USACE government employees, soldiers from the 249th Engineer Battalion, and contractors. USACE installed a record number of emergency electric generators in Puerto Rico—over 2,300—through the end of May 2018. The previous record was 310 emergency generators installed in response to Hurricane Katrina. On May 17, 2018, FEMA approved the extension of the USACE mission assignment for emergency power to November 30, 2018. This extension permitted USACE to continue its support for the more than 700 generators still in use throughout Puerto Rico at that time. FEMA later extended the mission assignment until April 8, 2019. As of December 11, 2018, USACE was supporting 24 generators in Puerto Rico, seven of which were supporting micro grids for the island municipalities of Vieques and Culebra. In the U.S. Virgin Islands, USACE installed 180 generators as a part of its temporary emergency power mission. USACE's temporary emergency power mission for the U.S. Virgin Islands was completed in February 2018, and USACE is no longer supporting generators there.

Unprecedented Roles by FEMA and USACE in Puerto Rico

In addition to the typical roles federal agencies undertake in restoration activities, FEMA and USACE undertook unprecedented roles in Puerto Rico because of the severe and widespread impacts of Hurricane Maria and PREPA's limited capacity. For the first time in its history, FEMA undertook the role of helping to coordinate major electricity grid

---

[19] In Puerto Rico, FEMA also assigned USACE the task of maintaining and fueling hundreds of other nonfederal generators. Similar to its mission in Puerto Rico, USACE provided support in the U.S. Virgin Islands by installing generators for critical facilities, but FEMA contractors maintained and fueled those generators.

restoration because PREPA did not have the necessary capability, capacity, or structure to respond, according to FEMA officials. FEMA officials also noted that PREPA's workers were not only engaged in restoration work but were also victims dealing with the same post-hurricane effects as the rest of the population.

As part of its response, FEMA mission assigned USACE to lead federal efforts to repair Puerto Rico's electricity grid—a role USACE had not played in the past during a domestic disaster response. Specifically, on September 30, 2017, the FEMA Administrator tasked USACE with leading the planning, coordination, and integration of the grid restoration. FEMA assigned USACE to lead federal efforts and provide direct support for grid restoration because PREPA was overwhelmed and had liquidity issues and USACE had the structures in place to award contracts with and bring in grid restoration crews, according to FEMA officials. In order to carry out its mission assignment, USACE issued contracts to bring lineworkers and materials to Puerto Rico to support the reinstallation and repair of transmission and distribution lines, among other power restoration activities.[20] As of June 30, 2018, USACE had obligated approximately $1.5 billion on these contracts. Figure 2 shows USACE and its contractors working to restore electricity in Puerto Rico. USACE's grid restoration mission assignment from FEMA ended on May 18, 2018, because, according to FEMA officials, power had been restored to approximately 98 percent of customers and PREPA, with its remaining contractors, had adequate capability to do the remaining work.

In addition to the federal response, PREPA issued its own contracts to bring in additional lineworkers, received assistance from the New York State Utility Contingent, and requested and received mutual assistance from other utilities.[21]

---

[20] Specifically, USACE issued contracts for the labor, equipment, materials, and supervision required to restore electrical transmission and distribution systems; to repair electrical generation systems, such as existing PREPA power plants; and to provide temporary generators.

[21] Specifically, the Governor of Puerto Rico requested assistance from the Governor of New York through the Emergency Management Assistance Compact—a mutual aid agreement among member states that provides a legal framework for requesting resources. New York utilities established the New York State Utility Contingent to organize and deploy resources

Sources: (from left to right) Robert DeDeaux, USACE, Engineer Research and Development Center; Dave Palmer, USACE, Los Angeles District; USACE, Tulsa District. | GAO-19-296.

Figure 2. U.S. Army Corps of Engineers (USACE) and USACE Contractors Working to Restore Electricity in Puerto Rico.

PREPA did not initially reach out for mutual assistance. About 6 weeks following Hurricane Maria, on October 31, 2017, PREPA formally requested aid from other utilities on the mainland through the American Public Power Association and the Edison Electric Institute. The electric power industry sent two individuals to Puerto Rico on November 3, 2017 and they began assessing storm damage and working with PREPA, FEMA, USACE, and DOE officials to develop a restoration plan. On November 22, 2017, the Governor of Puerto Rico appointed one of these individuals as Power Restoration Coordinator to oversee the multipronged restoration effort. According to the Power Restoration Coordinator, as a first step he worked to create an incident command structure, and incident management teams began arriving in December. Once the incident command structure was in place, the industry deployed additional crews, equipment and materials in January to accelerate the ongoing restoration efforts across the island. As discussed previously, local utilities are typically responsible for

to Puerto Rico. The Office of the Governor of New York provided leadership, and the New York Power Authority and Consolidated Edison Company of New York, Inc. coordinated the contingent. According to the New York State Utility Contingent's after action report, using the Emergency Management Assistance Compact rather than traditional utility mutual assistance required a unique set of processes, procedures, financial accounting, and restrictions to which the utilities are traditionally not accustomed. See New York State Utility Contingent, *Emergency Response to Hurricane Maria: After Action Report* (White Plains, N.Y.: August 2018).

restoring service, with federal agencies providing financial and other support. In contrast, approximately half of the lineworkers working to restore the electricity grid in Puerto Rico were USACE or USACE contractors at the peak of restoration efforts in February 2018, as shown in figure 3.

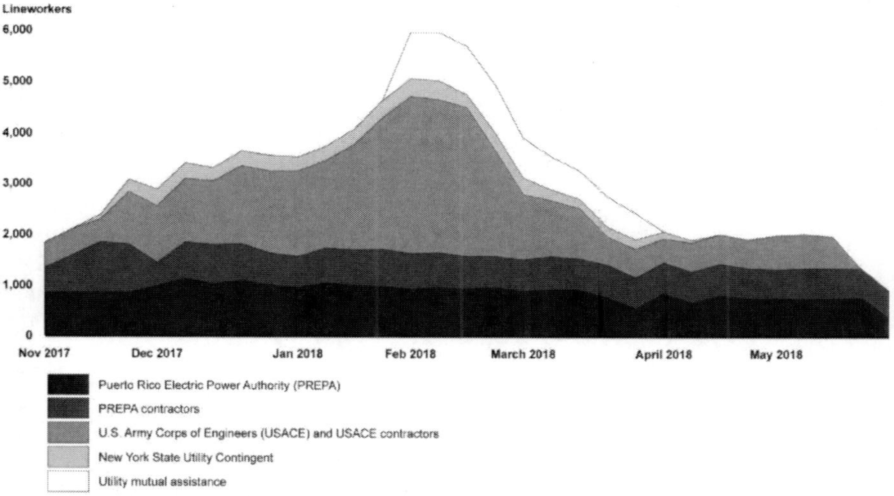

Source: GAO analysis of PREPA information | GAO-19-296.

Note: Electric utilities often use mutual assistance—voluntary partnerships with other electric utilities—to bring in additional resources beyond those of the affected utility to help restore electricity. New York utilities established the New York State Utility Contingent to organize and deploy resources to Puerto Rico. The Office of the Governor of New York provided leadership, and the New York Power Authority and Consolidated Edison Company of New York, Inc. coordinated the contingent.

Figure 3. Number of Lineworkers Who Worked to Restore the Electricity Grid in Puerto Rico from November 2017 through May 2018.

FEMA established a unified command structure to coordinate efforts of federal agencies, PREPA, PREPA's contractors, the New York State Utility Contingent, and utilities providing mutual assistance to PREPA, to help target priority work, ensure that crews could get to the work, and identify needed materials. Figure 4 shows the unified command structure.

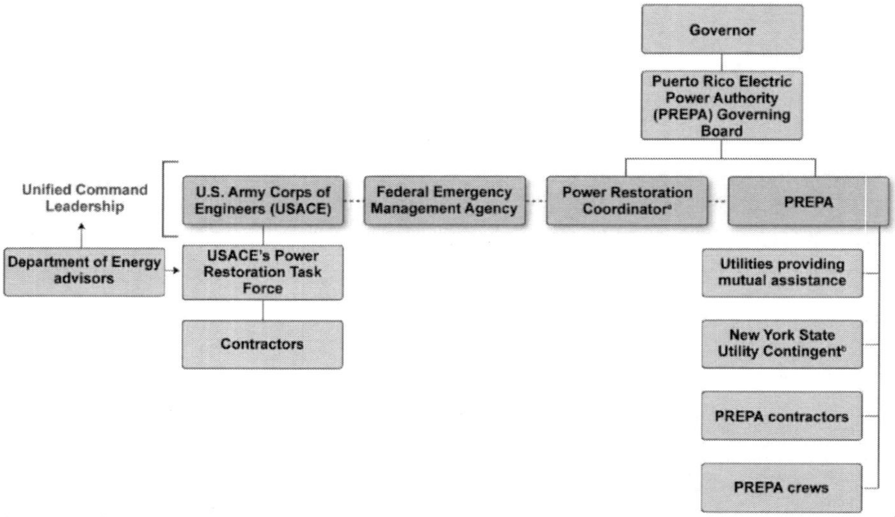

Source: GAO analysis of Federal Emergency Management Agency information | GAO-19-296.

[a] The Governor of Puerto Rico appointed a Power Restoration Coordinator to oversee the multipronged restoration effort.

[b] New York utilities established the New York State Utility Contingent to organize and deploy resources to Puerto Rico. The Office of the Governor of New York provided leadership, and the New York Power Authority and Consolidated Edison Company of New York, Inc. coordinated the contingent.

Figure 4. Unified Command Structure for Electricity Grid Restoration in Puerto Rico.

Logistical Challenges and Other Factors Affected Federal Support to Restore Electricity

According to documents we reviewed and our interviews with officials and representatives, the most commonly cited factors that affected federal electricity grid restoration efforts in Puerto Rico and the U.S. Virgin Islands included (1) logistical challenges, (2) availability of materials, (3) financial condition of local utilities and poor condition of existing infrastructure, and (4) the extensive and unprecedented role of federal agencies

- *Logistical challenges.* Responding to disasters on islands presents a number of logistical challenges. Specifically, according to federal officials, getting the crews, equipment, and materials needed to support restoration efforts to an island was more difficult and time-consuming than doing so on the mainland. This includes prepositioning assets, such as generators, and delivering equipment and materials in advance of a storm. The difficulties were greater in the days following the hurricanes since neither the ports nor the airports in Puerto Rico and the U.S. Virgin Islands had power, which prevented the delivery of materials to the islands. In Puerto Rico, the Port of San Juan reopened for daylight operations 3 days after Hurricane Maria made landfall; every airport and seaport had limited capacity after reopening for approximately 7 days post-landfall, according to FEMA's 2017 Hurricane Season After-Action Report.[22] Federal officials in the U.S. Virgin Islands told us that they faced further delays locating key supplies because of inadequate labelling of shipping containers at the port. For example, some containers were marked only as disaster supply equipment, which did not sufficiently identify the contents within them. According to USACE's 2018 Remedial Action Program Senior Leader Briefing, USACE lacked the expertise and capabilities to manage the large operational logistics requirements to support the Puerto Rico and U.S. Virgin Islands response.[23]
- *Availability of materials.* The sequence of three hurricanes making landfall in the United States in 2017 and the need to restore electricity service in Texas, Florida, and elsewhere, in addition to Puerto Rico and the U.S. Virgin Islands, complicated the restoration effort in the two territories. Since utilities in all affected areas were acquiring materials to restore electricity service, demand for these materials increased and available supplies were generally low; in

---

[22] Federal Emergency Management Agency, *2017 Hurricane Season FEMA After-Action Report* (Washington, D.C.: July 12, 2018). The report identifies lessons learned from the 2017 hurricane season to target improvements in FEMA's response and recovery operations.

[23] U.S. Army Corps of Engineers, Remedial Action Program Senior Leader Briefing (July 19, 2018).

some instances materials were only available as they were manufactured. Few, if any, materials were stockpiled locally on Puerto Rico. In addition, some of the equipment used in Puerto Rico was not standard in the continental United States and required ordering of specialized materials, resulting in delays in the restoration process. The U.S. Virgin Islands also faced supply issues, which became worse once grid recovery work in Puerto Rico began.

- *Financial condition of local utilities and poor condition of existing infrastructure.* Electric utilities in both Puerto Rico and the U.S. Virgin Islands were insolvent, which led to a lack of maintenance and presented its own challenges for restoring the grids after the storms. Specifically, PREPA was approximately $9 billion in debt before the 2017 hurricane season, with annual costs that exceeded its revenues. Puerto Rico's electric power infrastructure was in poor condition before the 2017 hurricane season largely because of PREPA's underinvestment and poor maintenance practices. For example, PREPA canceled its vegetation management program because of its financial situation; this contributed to the destruction of transmission and distribution lines when the hurricane arrived, according to FEMA officials.

  Similarly, in the U.S. Virgin Islands, financial challenges contributed to the extent of the damage to grid infrastructure. Specifically, VIWAPA officials told us that VIWAPA's financial challenges prevented certain infrastructure improvements and a large proportion of the electricity poles were at or above their weight capacity, increasing the likelihood and extent of wind damage during the hurricanes. According to VIWAPA officials, VIWAPA was aware that there were too many lines and heavy transformers on old poles, but was not in a position to address this concern prior to the hurricanes.
- *Extensive and unprecedented role of federal agencies.* FEMA did not anticipate or plan for the extensive role that it and USACE played in grid restoration in Puerto Rico. According to FEMA's

after action report for the 2017 hurricane season, FEMA's planning assumptions for a hurricane, earthquake, or tsunami striking Puerto Rico and the U.S. Virgin Islands underestimated the actual requirements.[24] As discussed above, prior to Hurricane Maria in Puerto Rico, USACE had never worked on a large-scale power restoration as part of a domestic disaster response and did not have expertise in this area, according to USACE officials. This affected grid restoration efforts. For example, USACE did not have a grid restoration contract in place to immediately initiate grid repair efforts, according to USACE officials. Rather, USACE issued an order off of a pre-existing contract that it had under its public works and engineering support function to bring electric utility lineworkers to Puerto Rico. According to USACE officials, a bid protest against one of USACE's contracts delayed its ability to increase the contract to bring more lineworkers to Puerto Rico.[25] In addition, the contract review and approval process USACE used to obtain supplies took longer than it would typically take utilities to get supplies, according to FEMA officials we interviewed.[26] According to USACE officials, USACE followed federal acquisition regulations, which is a slow process compared to private party purchases. USACE officials said that USACE is considering looking at what would be needed to create an advance grid restoration contract.

FEMA, USACE, and DOE identified potential actions to address these challenges. According to its after action report, FEMA plans to establish a standing interagency Power Task Force to coordinate with DOE, USACE, and state and local governments and provide crisis planning for the energy sector emergency support function to support the restoration of power

---

[24] Federal Emergency Management Agency, 2017 Hurricane Season FEMA After-Action Report.

[25] A bid protest is a challenge to the award or proposed award of a contract for the procurement of goods and services or a challenge to the terms of a solicitation for such a contract.

[26] According to USACE officials, contracting inefficiencies, such as changes in the scope of work led to some delays. We are conducting a review of 2017 post-disaster contracts, which we plan to issue in 2019.

during future national response efforts.[27] USACE's 2018 Remedial Action Program Senior Leader Briefing made recommendations to improve contingency contracting and operational logistics, among other things.[28] Specifically, recommendations included that USACE review existing and planned advance contracts and make adjustments as necessary to increase capacity and improve capabilities, and that USACE work with FEMA to convene an interagency logistics planning team and identify logistics contracting gaps and propose government and private sector solutions. DOE's after action report for the 2017 hurricane season says that the lessons learned from the response to Hurricane Maria may prompt some programmatic improvements to the energy sector emergency support function roles and responsibilities related to island response, among other potential improvements.[29] In addition, the report states that because of the extensive damage to grid infrastructure and the length of the restoration and recovery, there is an increasing need to incorporate resilience and hardening into restoration, recovery, and mitigation planning and execution.

AGENCY COMMENTS

We provided a draft of this chapter to the Department of Defense (DOD), the Department of Homeland Security (DHS), DOE, and the governments of Puerto Rico and the U.S. Virgin Islands for review and comment. In its comments, reproduced in appendix II, DHS indicated that a top priority of DHS, FEMA and the entire federal government has been to provide life safety and life-sustaining resources to Puerto Rico and the U.S. Virgin Islands, including efforts to restore power and stabilize critical infrastructure. DHS, DOD, and DOE also provided technical comments, which we incorporated as appropriate.

---

[27] Federal Emergency Management Agency, 2017 Hurricane Season FEMA After-Action Report.

[28] U.S. Army Corps of Engineers, Remedial Action Program Senior Leader Briefing (July 19, 2018).

Frank Rusco
Director, Natural Resources and Environment.

LIST OF REQUESTERS

The Honorable Michael Enzi
Chairman
Committee on the Budget
United States Senate

The Honorable Ron Johnson
Chairman

The Honorable Gary C. Peters
Ranking Member
Committee on Homeland Security and Governmental Affairs
United States Senate

The Honorable Rand Paul
Chairman
Subcommittee on Federal Spending Oversight and Emergency Management
Committee on Homeland Security and Governmental Affairs
United States Senate

The Honorable Marco Rubio
Chairman
Committee on Small Business and Entrepreneurship
United States Senate

The Honorable Maxine Waters
Chairwoman
Committee on Financial Services
House of Representatives

The Honorable Sean P. Duffy
Ranking Member
Subcommittee on Housing, Community Development and Insurance
Committee on Financial Services
House of Representatives

The Honorable Al Green
Chairman
Subcommittee on Oversight Investigations
Committee on Financial Services
House of Representatives

The Honorable Bennie G. Thompson
Chairman
Committee on Homeland Security
House of Representatives

The Honorable Elijah E. Cummings
Chairman

The Honorable Jim Jordan
Ranking Member
Committee on Oversight and Reform
House of Representatives

The Honorable Nydia Velázquez
Chairwoman
Committee on Small Business
House of Representatives

The Honorable Peter A. DeFazio
Chairman

The Honorable Sam Graves
Ranking Member
Committee on Transportation and Infrastructure
House of Representatives

The Honorable Emanuel Cleaver, II
House of Representatives

The Honorable Michael McCaul
House of Representatives

The Honorable Gary Palmer
House of Representatives

The Honorable Ann Wagner
House of Representatives

APPENDIX I: TIMELINES OF FEDERAL AND OTHER EFFORTS TO SUPPORT ELECTRICITY GRID RESTORATION

Figure 5. (Continued)

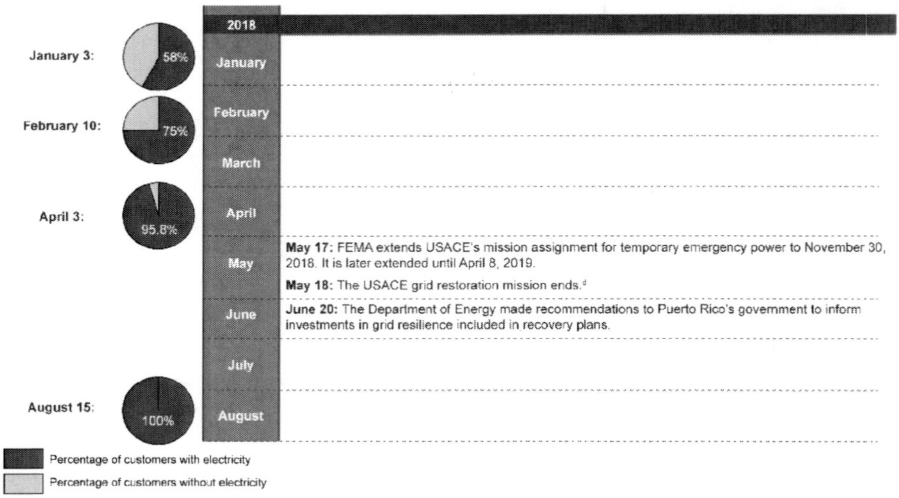

Source: GAO analysis of information from FEMA, PREPA, USACE, the Department of Energy, the New York State Utility Contingent, and the Office of Governor of Puerto Rico. | GAO-19-296.

[a] PREPA's estimates of customers with power restored are based on the number of meters that PREPA knows are served by a given power line and on the number of meters PREPA can read currently, according to a PREPA official. Power has been restored to 100 percent of customers that are able to safely receive power, but this does not mean that all prestorm customers have power restored, as some structures may not have been deemed safe for power restoration, according to PREPA officials.

[b] The declaration was amended on September 26, 2017 to authorize increased cost sharing to 100 percent for emergency protective measures beginning September 17, 2017, including direct federal assistance, for 180 days from the start of the incident period. The100 percent federal funding was later extended until August 16, 2018, for power restoration and until September 15, 2018, for emergency temporary power.

[c] For the purposes of this chapter, contract obligations include obligations against what the Federal Procurement Data System-Next Generation (FPDS-NG) categorizes as definitive vehicles (definitive contracts and purchase orders that have a defined scope of work that do not allow for individual orders under them) and against what FPDS-NG categorizes as indefinite delivery vehicles (including orders under indefinite delivery vehicles, such as indefinite delivery, indefinite quantity contracts).

[d] USACE will no longer provide line restoration work for PREPA, but FEMA extended USACE's logistics and materials management capability until PREPA can effectively manage the volume of emergency restoration materials.

Figure 5. Timeline of Federal and Other Efforts to Support Electricity Grid Restoration in Puerto Rico after the 2017 Hurricane Season.

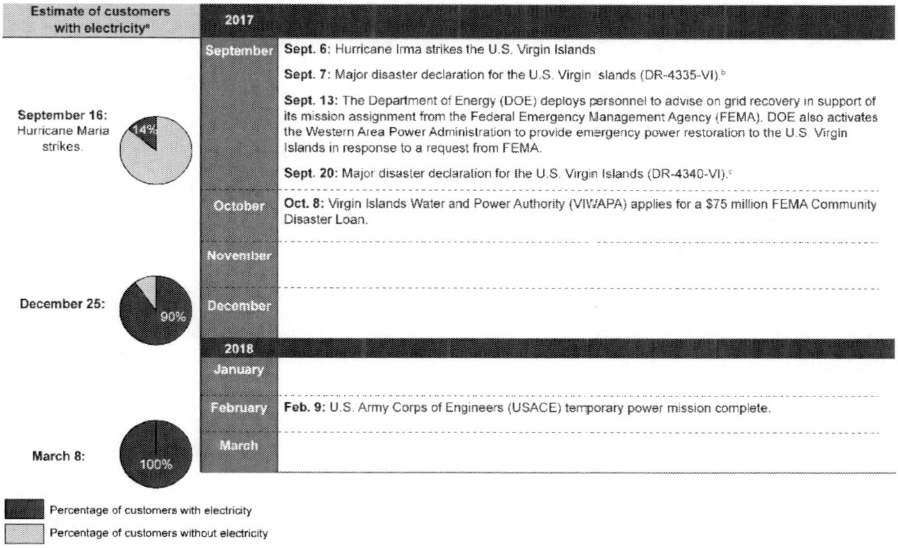

Source: GAO analysis of USACE, FEMA and VIWAPA documents. | GAO-19-296.

Note: VIWAPA contracted for assistance with restoring electricity and also received mutual assistance, according to VIWAPA officials.

[a] The share of customers with power restored is out of the total number of customers deemed safe for power restoration, according to VIWAPA. This does not mean that all prestorm customers have power, as some structures may not have been deemed safe for power restoration. VIWAPA's estimates of customers with power restored are based on the number of meters that VIWAPA knows are served by a given power line, as VIWAPA's automated system for determining the percentage of customers without power was destroyed and is still being restored, according to a FEMA official.

[b] The declaration was amended on September 28, 2017, to authorize increased cost sharing to 100 percent for emergency protective measures beginning September 6, including direct federal assistance, for 180 days from the start of the incident period and a 90 percent federal cost share thereafter. On March 15, 2018, the 100 percent federal funding for emergency protective measures was extended for 60 days, and on May 18, 2018, the 100 percent federal funding for emergency temporary power support was extended for 120 days.

[c] The declaration was amended on October 3, 2017, to authorize increased cost sharing to 100 percent for emergency protective measures beginning September 16, including direct federal assistance, for 180 days from the start of the incident period and a 90 percent federal cost share thereafter. On March 15, 2018, the 100 percent federal funding for emergency protective measures was extended for 60 days, and on May 18, 2018 the 100 percent federal funding for emergency temporary power support was extended for 120 days.

Figure 6. Timeline of Federal and Other Efforts to Support Electricity Grid Restoration in the U.S. Virgin Islands after the 2017 Hurricane Season.

See figures 5 and 6 for a timeline of federal and other efforts to support electricity grid restoration in Puerto Rico and the U.S. Virgin Islands after the 2017 hurricane season.

APPENDIX II: COMMENTS FROM THE DEPARTMENT OF HOMELAND SECURITY

U.S. Department of Homeland Security
Washington, DC 20528

Homeland Security

April 2, 2019

Chris P. Currie
Director, Homeland Security and Justice Issues
U.S. Government Accountability Office
441 G Street, NW
Washington, DC 20548

Re: Management Response to Draft Report GAO-19-296, "2017 HURRICANE SEASON: Federal Support for Electricity Grid Restoration in the U.S. Virgin Islands and Puerto Rico"

Dear Mr. Currie:

Thank you for the opportunity to review and comment on this draft report. The U.S. Department of Homeland Security (DHS) appreciates the U.S. Government Accountability Office's (GAO) work in planning and conducting its review and issuing this report.

The Department is pleased to note GAO's positive acknowledgement of the Federal Emergency Management Agency's (FEMA's) unprecedented role in coordinating and directly assisting with electricity grid restoration efforts in Puerto Rico and U.S. Virgin Islands (USVI) after Hurricanes Irma and Maria in 2017. As of July 20, 2018, FEMA had obligated $3.2 billion in Puerto Rico and $795 million in the USVI in both Public Assistance grant funds and direct federal assistance through mission assignments. A top priority of DHS, FEMA and the entire federal government has been continuing to provide life safety and life sustaining resources to Puerto Rico and USVI, including efforts to restore power and stabilize critical infrastructure.

In Puerto Rico, FEMA's 2019 Puerto Rico Contingency Plan includes the use of historical data to identify critical facilities that will need generators. To meet this requirement FEMA not only identified the number of generators needed, but also added generators with higher capacities. To improve resiliency of the electric grid in the USVI, the USVI Water and Power Authority (WAPA) is replacing wooden poles with composite poles which can withstand wind speeds of 200 miles per hour. USVI WAPA is also working on a plan to further improve resiliency by putting electric utility lines underground.

The National Response Framework (NRF) is part of the National Strategy for Homeland Security that presents the guiding principles enabling all levels of domestic response partners to prepare for and provide a unified national response to disasters and emergencies. The NRF is one of 5 frameworks established as part of the National Preparedness System which includes the National Frameworks for Prevention, Protection, Response, Recovery, and Mitigation as well as the National Preparedness Goal. The NRF designates FEMA as the lead coordinator of the 14 agencies supporting each function for federal response efforts. In building capabilities to restore Puerto Rico and USVI electricity grids, FEMA coordinated efforts with the Department of Energy (DOE), United States Army Corps of Engineers (USACE), Puerto Rico Electric Power Authority (PREPA), and USVI WAPA.

FEMA also formed the Emergency Power Task Force (TEPTF) to validate the stock and current status of the FEMA/Federal generators that have been stored in Puerto Rico. The TEPTF will coordinate with its Federal partners to initiate the installation of generators based on Puerto Rico's stated priorities, when needed. The TEPTF, which includes the DOE and USACE, meets quarterly to discuss topics related to interagency and private sector coordination, collaboration, and support. If needed, the TEPTF will evolve into a Crisis Action Planning Team to support FEMA's National Response Coordination Center and Regional Response Coordination Center activations.

The draft report did not contain any recommendations. Technical comments were previously provided under separate cover. FEMA remains committed to supporting our citizens and first responders before, during, and after disasters.

Again, thank you for the opportunity to review and comment on this draft report. We look forward to working with you again in the future. Please feel free to contact me if you have any questions.

Sincerely,

JIM H. CRUMPACKER, CIA, CFE
Director
Departmental GAO-OIG Liaison Office

In: Hurricanes, Wildfires and Flooding
Editor: Lydie Yohan
ISBN: 978-1-53616-362-9

Chapter 3

PUERTO RICO HURRICANES: STATUS OF FEMA FUNDING, OVERSIGHT, AND RECOVERY CHALLENGES*

United States Government Accountability Office

ABBREVIATIONS

BBA	Bipartisan Budget Act of 2018
COR3	Central Office for Recovery, Reconstruction and Resilience
DHS	Department of Homeland Security
DRF	Disaster Relief Fund
EMMIE	Emergency Management Mission Integrated Environment
FEMA	Federal Emergency Management Agency
FOMB	Financial Oversight and Management Board
GAR	Governor's Authorized Representative

* This is an edited, reformatted and augmented accessible version of the United States Government Accountability Office Report to Congressional Requesters, Publication No. GAO-19-256, dated March 2019.

HSOAC	Homeland Security Operational Analysis Center
IFMIS	Integrated Financial Management Information System
NOAA	National Oceanic and Atmospheric Administration
Stafford Act	Robert T. Stafford Disaster Relief and Emergency Assistance Act

WHY GAO DID THIS STUDY

In 2017 two major hurricanes – Irma and Maria – caused extensive damage throughout Puerto Rico. Hurricane Maria, a Category 4 hurricane, was the most intense hurricane to make landfall in Puerto Rico since 1928, destroying roads, buildings, and cutting power and communication lines, among other things. Puerto Rico estimates that $132 billion will be needed to repair and reconstruct infrastructure and services. FEMA—a component of the Department of Homeland Security (DHS)—is the lead federal agency responsible for assisting Puerto Rico as it recovers. FEMA administers the Public Assistance program in partnership with Puerto Rico to provide funds to rebuild damaged infrastructure and restore critically-needed services.

GAO was asked to review the federal government's recovery efforts related to the 2017 hurricanes. This chapter, among other objectives, describes (1) FEMA's Public Assistance spending in Puerto Rico and oversight efforts of federal recovery funds, and (2) initial challenges with the recovery process. GAO reviewed Public Assistance program documents; analyzed grant funding data; and interviewed officials from Puerto Rico and DHS about the Public Assistance program and recovery efforts, as well as officials from ten municipalities selected on the basis of population and Public Assistance spending.

GAO is not making recommendations at this time, but will continue monitoring the recovery as part of its ongoing work.

WHAT GAO FOUND

The Federal Emergency Management Agency (FEMA) obligated almost $4 billion in Public Assistance grant funding to Puerto Rico as of September 30, 2018 in response to the 2017 hurricanes. FEMA obligated about $3.63 billion for emergency work—emergency measures such as debris removal and generators—and about $151 million for permanent work to repair and replace public infrastructure such as roads (see figure).

Puerto Rico established a central recovery office to oversee federal recovery funds and is developing an internal controls plan to help ensure better management and accountability of the funds. In the interim, FEMA instituted a manual reimbursement process—requiring FEMA to review each reimbursement request before providing Public Assistance funds—to mitigate risk and help ensure financial accountability. FEMA officials stated that they will remove this manual process once the agency approves Puerto Rico's internal controls plan.

Officials from FEMA and Puerto Rico's central recovery office and municipalities that GAO interviewed reported initial challenges with the recovery process, including with Public Assistance alternative procedures. Unlike in the standard Public Assistance program where FEMA will fund the actual cost of a project, the Public Assistance alternative procedures allow awards for permanent work projects to be made on the basis of fixed cost estimates to provide financial incentives for the timely and cost-effective completion of work. Challenges identified included concerns about lack of experience and knowledge of the alternative procedures being applied in Puerto Rico; concerns about missing, incomplete, or conflicting guidance on the alternative procedures; and concerns that municipalities have not been fully reimbursed for work already completed in the immediate aftermath of the hurricanes, causing financial hardships in some municipalities.

FEMA officials stated that the agency is taking actions to address reported recovery challenges, such as leveraging existing expertise to train personnel and developing supplemental guidance on alternative procedures and reducing delays in reimbursements. GAO will continue to monitor these issues and plans to report additional findings and recommendations as appropriate later this year.

March 14, 2019
Congressional Requesters

On September 6, 2017, Hurricane Irma's eyewall passed 50 nautical miles north of Puerto Rico as a category 5 hurricane, causing severe wind and rain inundation to the main island and significant damage to the island of Culebra. Less than two weeks later, on the morning of September 20, 2017, Hurricane Maria made a direct hit as a category 4 storm on the main island of Puerto Rico.[1] Although the storm moved offshore in the early afternoon of September 20, 2017, tropical storm-force winds continued into the late night hours, causing extensive damage to electrical utilities, roads, bridges, and other public infrastructure throughout Puerto Rico. According to the National Hurricane Center, Hurricane Maria was the most intense hurricane to make landfall in Puerto Rico since 1928. In its recovery plan, Puerto Rico estimates that $132 billion in funding from 2018 through 2028 will be needed to repair and reconstruct the infrastructure damaged by the hurricanes. This includes $30 billion to comprehensively modernize Puerto Rico's energy sector and improve its resiliency, nearly $5 billion to replace Puerto Rico's drinking water system, and between $8 to $12 billion to rehabilitate owner-occupied homes that were substantially damaged by the hurricanes, among other things.[2]

The Federal Emergency Management Agency (FEMA), a component of the Department of Homeland Security (DHS), leads the federal effort to mitigate, respond to, and recover from disasters, both natural and

---

[1] The National Oceanic and Atmospheric Administration (NOAA) measures hurricanes on a scale from 1 to 5 with a Category 1 being the least intense and a Category 5 being the most intense. NOAA defines a Category 4 hurricane as one with winds above 130 miles per hour but less than 157 miles per hour. Hurricane Maria made landfall on Puerto Rico as a Category 4 hurricane.

[2] Central Office for Recovery, Reconstruction and Resiliency, Government of Puerto Rico, *Transformation and Innovation in the Wake of Devastation: An Economic and Disaster Recovery Plan for Puerto Rico* (August 8, 2018).

manmade.[3] Among other responsibilities, FEMA administers the Public Assistance program in partnership with the Government of Puerto Rico (Puerto Rico), providing grant funding for a wide range of response and recovery activities. These activities include, among other efforts, debris removal; life-saving emergency protective measures; and the repair, replacement, or restoration of disaster-damaged publicly-owned facilities. In addition, FEMA coordinated with Puerto Rico to submit an economic and disaster recovery plan to Congress, as required by the Bipartisan Budget Act of 2018.[4] FEMA also plans to work with Puerto Rico to ensure that it has effective internal controls in place to oversee federal recovery funds.

You asked us to review the federal government's response and recovery efforts related to the 2017 hurricanes. As part of that effort, we have several reviews ongoing of FEMA's and Puerto Rico's disaster recovery planning, oversight, and execution efforts in Puerto Rico. This chapter provides information on (1) FEMA Public Assistance program spending and actions FEMA and Puerto Rico have taken to provide oversight of federal recovery funds, (2) initial challenges identified by officials from FEMA and Puerto Rico with the recovery process, and (3) Puerto Rico's economic and disaster recovery plan.

We have another ongoing review that will examine disaster recovery efforts in Puerto Rico and we plan to issue that report in 2019. In addition, we are conducting a broader body of work covering various disaster response and recovery issues, including disaster recovery in the U.S. Virgin Islands and impacted states, and will also be reporting on those issues over the next year.

For this chapter, we reviewed relevant FEMA Public Assistance program documentation, and we obtained and analyzed data from FEMA's Emergency Management Mission Integrated Environment (EMMIE) and Integrated Financial Management Information System (IFMIS) databases. These databases include data on Public Assistance program obligations and

[3] See 6 U.S.C. § 313

[4] Pub. L. No. 115- 123, § 21210, 132 Stat. 64, 108-09 (2018). Section 21210 requires that Puerto Rico submit to Congress a report describing Puerto Rico's 12- and 24-month economic and disaster recovery plan not later than 180 days after the date of enactment, February 9, 2018.

expenditures related to Hurricanes Irma and Maria in Puerto Rico as of the end of fiscal year 2018.[5] We assessed the reliability of EMMIE and IFMIS data by reviewing existing information about these systems, interviewing data users and managers responsible for these data, and cross-checking data across disparate sources to ensure consistency. Based on these steps, we determined these data to be sufficiently reliable for the purposes of describing Public Assistance program obligations and expenditures for Hurricanes Irma and Maria in Puerto Rico as of the end of fiscal year 2018. An obligation is a definite commitment that creates a legal liability of the government for the payment of goods and services ordered or received. An expenditure is an amount paid by federal agencies, by cash or cash equivalent, during the fiscal year to liquidate government obligations. For the purpose of this chapter, an expenditure represents the actual spending by Puerto Rico of money obligated by the federal government.

We interviewed officials from FEMA headquarters, as well as officials from FEMA Region II, the FEMA regional office with oversight for Puerto Rico. Further, in May, August, and September of 2018, we conducted site visits to Puerto Rico to observe hurricane-damaged areas and interview relevant officials. In Puerto Rico, we interviewed FEMA officials at the joint recovery office, a multiagency coordination center established in Guaynabo, Puerto Rico to coordinate major disaster recovery efforts. At this location, we interviewed the federal coordinating officer for Puerto Rico, officials from the Public Assistance program, and officials from FEMA's finance and administration section to understand FEMA's recovery operations in Puerto Rico. In addition, we interviewed Puerto Rico government officials from the Central Office for Recovery, Reconstruction and Resilience (COR3)[6] and ten selected municipal

---

[5] These data do not include obligations and expenditures for direct federal mission assignments, in which a federal agency is tasked with providing eligible emergency work or debris removal services to a territory or state, or for other categories of mission assignments.

[6] As outlined in the Governor's order establishing COR3, COR3 will identify, procure, and administer all federal, territory, and private resources available to Puerto Rico related to recovery. Additionally, COR3's Executive Director will act as the Governor's Authorized Representative (GAR). The GAR is the designated individual responsible for administering federal disaster assistance programs on behalf of the state, local, tribal, or territorial

governments throughout Puerto Rico.[7] The results from our interviews cannot be generalized to all of Puerto Rico's 78 municipalities; however, they provided important context about the status of recovery efforts in various locations throughout Puerto Rico. We spoke with officials from the Financial Oversight and Management Board (FOMB), a board established in the Puerto Rico Oversight, Management, and Economic Stability Act with broad budgetary and financial control over Puerto Rico.[8] We also interviewed the comptroller of Puerto Rico to obtain her input on Puerto Rico's government structures and municipal issues. Lastly, we reviewed Puerto Rico government documents related to recovery planning and internal controls.

We conducted this performance audit from March 2018 to March 2019 in accordance with generally accepted government auditing standards. Those standards require that we plan and perform the audit to obtain sufficient, appropriate evidence to provide a reasonable basis for our findings and conclusions based on our audit objectives. We believe that the evidence obtained provides a reasonable basis for our findings and conclusions based on our audit objectives.

---

government and ensuring compliance with all applicable laws, regulations, and FEMA guidance.

[7] During an August 2018 trip to Puerto Rico, we met with officials from the municipalities of Canóvanas, Cayey, Orocovis, and Toa Baja to gather preliminary information and to develop the semi-structured interviews that took place in September 2018. During the September 2018 visit, we conducted semi-structured interviews with Puerto Rico's six most populous municipalities and the four municipalities with the highest level of Public Assistance spending per capita in each of FEMA's four administrative branches in Puerto Rico. According to 2017 Census estimates, the six most populous municipalities are, in order of population: San Juan, Bayamón, Carolina, Ponce, Caguas, and Guaynabo. According to FEMA, the four municipalities with the highest level of Public Assistance spending per capita per FEMA branch as of August 6, 2018 are: Maricao (Branch I), Naranjito (Branch II), Maunabo (Branch III), and Guayanilla (Branch IV).

[8] The Puerto Rico Oversight, Management, and Economic Stability Act established the Financial Oversight and Management Board (FOMB) to provide guidance to Puerto Rico in fiscal responsibility. See Pub. L. No. 114-187, 130 Stat. 549 (2016). FOMB is responsible for certifying Puerto Rico's financial plans, approving and monitoring budgets and budgetary activities, advising Puerto Rico on financial management, and certifying restructuring and approving actions related to debt issuance. FOMB is also responsible for certifying Puerto Rico's economic and disaster recovery plan, released on August 8, 2018.

BACKGROUND

On September 6, 2017, the eye of Hurricane Irma traveled about 50 nautical miles to the north of the northern shore of Puerto Rico as a category 5 hurricane. Less than two weeks later, Hurricane Maria made landfall as a category 4 hurricane on the main island of Puerto Rico on the morning of September 20, 2017 with wind speeds up to 155 miles per hour. The center of the hurricane moved through southeastern Puerto Rico to the northwest part of the island, as shown in figure 1 below.

In response to the request of the Governor of Puerto Rico, the President declared a major disaster the day after each hurricane impacted Puerto Rico.[9] Major disaster declarations can trigger a variety of federal response and recovery programs for government and nongovernmental entities, households, and individuals, including assistance through the Public Assistance program.[10] Under the National Response Framework, DHS is the federal department with primary responsibility for coordinating disaster response, and within DHS, FEMA has lead responsibility. The Administrator of FEMA serves as the principal adviser to the President and the Secretary of Homeland Security regarding emergency management.[11]

---

[9] In accordance with the Robert T. Stafford Disaster Relief and Emergency Assistance Act (Stafford Act), as amended, the President of the United States may declare that a major disaster or emergency exists in response to a Governor's or tribal chief executive's request if the disaster is of such severity and magnitude that effective response is beyond the capabilities of a state, tribe, or local government and federal assistance is necessary. See 42 U.S.C. § 5170. Prior to a major disaster or emergency declaration, a Preliminary Damage Assessment is typically used to determine the impact and magnitude of damage and the resulting needs of individuals, businesses, the public sector, and the community as a whole. The territory uses the information collected as the basis for the Governor's request for a major disaster or emergency declaration. However, based on the severity and magnitude of Hurricanes Irma and Maria, the FEMA Regional Administrator upon consultation with Puerto Rico, determined supplemental federal assistance was necessary before joint federal, state, and local government Preliminary Damage Assessments were completed.

[10] See 42 U.S.C. § 5170. Presidential Policy Directive-8: National Preparedness establishes a national preparedness system made of an integrated set of guidance, programs, and processes designed to strengthen the security and resilience of the United States through systematic preparation for the natural and human-caused threats that pose the greatest risk. This system breaks preparedness activities into five different lines of effort—prevention, protection, mitigation, response, and recovery—each of which requires a separate planning framework.

[11] 6 U.S.C. § 313(c)(4).

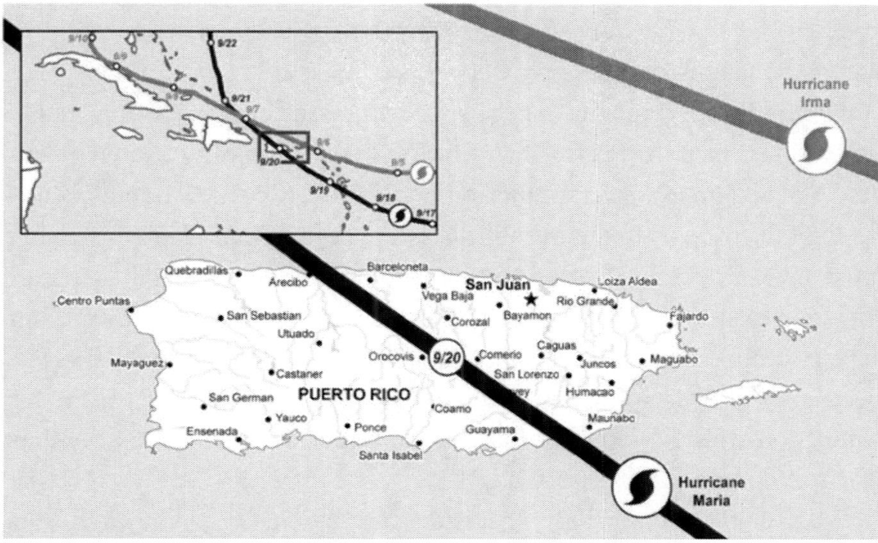

Source: GAO analysis of National Oceanic and Atmospheric Administration Map Resources. (map). | GAO- 19-256.

Figure 1. The Paths of Hurricanes Irma and Maria.

FEMA's Public Assistance Program

FEMA's Public Assistance program provides funding to state, territorial, local, and tribal governments to assist them in responding to and recovering from major disasters or emergencies.[12] As shown in figure 2, Public Assistance program funds are categorized broadly as either "emergency work" or "permanent work." Within those two broad categories are separate sub-categories. In addition to the emergency work and permanent work categories, FEMA's Public Assistance program includes category Z, which represents indirect costs, administrative

[12] The Public Assistance program represents the largest share of the Disaster Relief Fund (DRF), which is the primary source of federal disaster assistance for state and local governments when a disaster is declared. The DRF is appropriated no-year funding, which allows FEMA to fund, direct, coordinate, and manage response and recovery efforts—including certain efforts by other federal agencies and state and local governments, among others—associated with domestic disasters and emergencies.

expenses, and other expenses a recipient or subrecipient incurs in administering and managing projects.

Puerto Rico's agencies, such as the Department of Housing; public corporations, such the Puerto Rico Electric Power Authority and the Puerto Rico Aqueduct and Sewer Authority; and Puerto Rico's 78 municipalities are eligible to apply for the Public Assistance program. FEMA's Public Assistance program also provides funding for cost-effective hazard mitigation measures to reduce or eliminate the long-term risk to people and property from future natural and man-made disasters and their effects.[13] Specifically, FEMA provides funding for hazard mitigation measures in conjunction with the repair of disaster-damaged facilities to enhance their resilience during future disasters.

For example, a community that had a fire station damaged by a disaster could use Public Assistance funding to repair the facility and incorporate additional measures such as installing hurricane shutters over the windows to mitigate the potential for future damage.

In Puerto Rico, the Public Assistance program is administered through a partnership between FEMA and the recipient (Puerto Rico), which provides funding to eligible subrecipients (local or territory-level entities). Under the standard Public Assistance program process, once the President has declared a disaster, Public Assistance staff work with the recipient or subrecipients to help them document damages, identify eligible costs and work, and prepare requests for Public Assistance grant funds by developing project proposals. Officials then review and obtain approval of projects prior to FEMA obligating funds to reimburse recipients or subrecipients for eligible work.

[13] FEMA may fund hazard mitigation measures related to the damaged facilities receiving Public Assistance funding pursuant to section 406 of the Stafford Act, as amended. 42 U.S.C. § 5172; 44 C.F.R. § 206.226. FEMA Public Assistance policy allows three different methods to test for cost-effective hazard mitigation. First, under the 15 percent rule, hazard mitigation measures may amount up to 15 percent of the total eligible cost of repair work on a project. Second, certain hazard mitigation measures that have been predetermined to be cost-effective may qualify under the 100 percent rule, which permits the hazard mitigation as long as it does not exceed 100 percent of the eligible cost of the repair work on a project. Third, for measures that exceed eligible costs, the territory must demonstrate that the measure is cost effective through an acceptable benefit/cost analysis methodology.

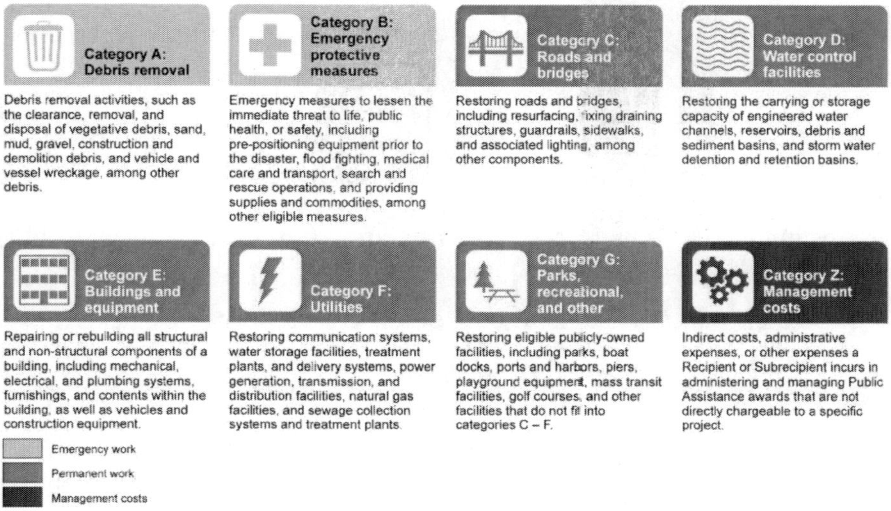

Source: Federal Emergency Management Agency. | GAO-19-256.

Figure 2. Federal Emergency Management Agency (FEMA's) Public Assistance Program Categories of Work.

The Use of Alternative Procedures for Public Assistance Projects in Puerto Rico

The Sandy Recovery Improvement Act of 2013 authorized the use of alternative procedures in administering the Public Assistance program, thereby providing new flexibilities to FEMA, states, territories, and local governments for debris removal, infrastructure repair, and rebuilding projects using funds from this program.[14] The stated goals of the alternative procedures are to reduce the costs to the federal government, increase flexibility in the administration of the Public Assistance program, expedite the provision of assistance under the program, and provide

[14] The Sandy Recovery Improvement Act of 2013 amended the Stafford Act by adding Section 428, which authorized FEMA to approve Public Assistance program projects under the alternative procedures provided by that section for any presidentially-declared major disaster or emergency. This section further authorized FEMA to carry out the alternative procedures as a pilot program until FEMA promulgates regulations to implement this section. Pub. L. No. 113-2, div. B, § 1102(2), 127 Stat. 39, amending Pub. L. No. 93-288, tit. IV, § 428 (codified at 42 U.S.C. § 5189f).

financial incentives for recipients of the program for the timely and cost-effective completion of projects.[15]

Unlike the standard Public Assistance program where FEMA will fund the actual cost of a project, the Public Assistance alternative procedures allow awards for permanent work projects to be made on the basis of fixed cost estimates to provide financial incentives for the timely and cost-effective completion of work.[16] Under these procedures, if the actual cost of the project exceeds the fixed cost estimate agreed upon by FEMA and the recipient, the recipient or subrecipient is responsible for the additional costs. However, if the actual cost of completing eligible work for a project is below the estimate, the recipient or subrecipient may use the remaining funds for other eligible purposes, such as for additional cost-effective hazard mitigation measures to increase the resiliency of public infrastructure. These funds may also be used for activities that improve the recipient's or subrecipient's future Public Assistance operations or planning. Although FEMA had approved alternative procedure grants in 30 states as of April, 2018, in these cases, alternative procedures were used on a project-by-project basis. Puerto Rico's recovery from the 2017 hurricanes is the first recovery to use alternative procedures for all large permanent work projects.

On October 30, 2017, Puerto Rico requested to use the alternative procedures process for all large-project funding for Public Assistance permanent work, categories C through G.[17] According to FEMA guidance, as part of the alternative procedures process in Puerto Rico, FEMA and Puerto Rico must agree on a group of personnel with cost estimation expertise who will serve as part of a center of excellence. This center of excellence will assist FEMA and Puerto Rico in developing cost estimating

[15] See 42 U.S.C. § 5189f(c).

[16] Public Assistance Alternative Procedures (Section 428) Guide for Permanent Work FEMA-4339-DR-PR (April, 2018).

[17] According to a November 2017 amendment to Puerto Rico's major disaster declaration, due to the extraordinary level of infrastructure damage caused by Hurricane Maria, as well as the financial status of Puerto Rico, officials chose to use the alternative procedures for all large-project funding for Public Assistance categories C through G pursuant to section 428 of the Stafford Act. Puerto Rico; Amendment No. 5 to Notice of a Major Disaster Declaration, 82 Fed. Reg. 53,514 (Nov. 16, 2017). For fiscal year 2018, the large project threshold is $125,500.

methodologies to be used for determining fixed cost estimates for Public Assistance permanent work projects. FEMA officials stated that they are in the process of conducting inspections for Public Assistance projects for permanent work and, as of August 2018, had a total list of 10,000 site inspections to complete. FEMA officials stated that October 2019 is their target date for completing all alternative procedures fixed cost estimates for Public Assistance permanent work.[18] However, pursuant to 428 guidance published in April 2018, this time frame may be adjusted on a project-by-project basis, based on extenuating circumstances.

Puerto Rico's Central Recovery Office Oversees Federal Recovery Spending

Amendment 5 to the President's disaster declaration imposed a number of grant conditions, including that Puerto Rico establish an oversight authority supported by third-party experts.[19] This authority is to act as the grant recipient for all Public Assistance and hazard mitigation funding to ensure sound project management and enhanced, centralized oversight over FEMA grant distributions. In October 2017, the Governor of Puerto Rico established COR3, a Puerto Rico government office, to plan, guide, and oversee recovery efforts, including administering and overseeing the Public Assistance program.[20] According to FEMA and COR3 officials, COR3 will fulfill the oversight requirements outlined in Amendment 5. According to COR 3 officials, COR3 was also established to ensure coordination with FEMA. The Executive Director of COR3 will act as the

---

[18] Amendment 5 to the notice of Puerto Rico's major disaster declaration also directs that estimates for projects exceeding a threshold amount are to be validated by an independent, third-party expert. Puerto Rico; Amendment No. 5 to Notice of a Major Disaster Declaration, 82 Fed. Reg. 53,514 (Nov. 16, 2017). FEMA has determined that this amount will be $5 million.

[19] 82 Fed. Reg. 53,514.

[20] COR3 has entered into contracts with four management consulting firms (CGI, Deloitte, ICF, and Navigant) to work on various projects including developing a recovery management plan and internal controls plan, and creating a disaster recovery information web portal, among others. According to COR3, these firms were retained using an open and competitive bid process.

Governor's Authorized Representative, which is the designated individual responsible for administering federal disaster assistance programs on behalf of Puerto Rico.[21] Among other things, COR3 will

- Identify, procure, and administer all federal, territorial, and private resources available to Puerto Rico related to recovery;
- Provide oversight of subrecipients using risk-based monitoring; and
- Provide technical assistance and advise Puerto Rico's governmental agencies and municipalities regarding any matter related to recovery.
- According to COR3 officials, they will also implement internal controls, policies, and procedures to appropriately manage recovery funds.

COR3 has also launched an online transparency portal intended to provide a breakdown of FEMA Public Assistance and other federal funding made available for disaster recovery in Puerto Rico.[22]

Bipartisan Budget Act of 2018 Requirements for Puerto Rico and Congressional Oversight of Recovery Efforts

The Bipartisan Budget Act of 2018 (Bipartisan Budget Act) required that Puerto Rico submit an economic and disaster recovery plan to Congress by August 9, 2018, that defines the priorities, goals, and expected outcomes of Puerto Rico's recovery related to a number of sectors, including, among other things, infrastructure, housing, electric power

---

[21] 44 C.F.R. §§ 206.2(a)(13) and 206.41(d). The Governor's Authorized Representative is the designated individual responsible for administering federal disaster assistance programs on behalf of the state, local, tribal, or territorial government and ensuring compliance with all applicable laws, regulations, and FEMA guidance.

[22] https://www.recovery.pr/.

systems and grid restoration.[23] The Bipartisan Budget Act also directs the Governor of Puerto Rico to develop the disaster recovery plan in coordination with FEMA, with support and contributions from other federal agencies having designated responsibilities in the National Disaster Recovery Framework.

As of June 2015, Puerto Rico had roughly $66.9 billion in outstanding debt. According to the recovery plan, economic contraction in the years prior to the hurricanes contributed to a severe fiscal crisis and Puerto Rico's credit rating dropped below investment grade in early 2014, followed by a series of defaults on debt payments. In response to Puerto Rico's financial crisis, Congress passed and the President signed the Puerto Rico Oversight, Management, and Economic Stability Act in June 2016, which established the FOMB with broad budgetary and financial control over Puerto Rico.[24] The Bipartisan Budget Act also permits the FOMB to review any federal funds over $10 million that are designated for Puerto Rico's response to or recovery from Hurricanes Irma or Maria.[25]

FEMA Obligated Almost $4 Billion in Public Assistance Funding, and Puerto Rico and FEMA Have Taken Actions to Provide Oversight of Federal Recovery Funds

FEMA obligated nearly $4 billion in Public Assistance funds for Puerto Rico's emergency work projects, as well as the repair and restoration of its public infrastructure, among other things. In order to

---

[23] [23]Pub. L. No. 115- 123, § 21210, 132 Stat. 64, 108-09 (2018). Other sectors specified in the Bipartisan Budget Act are: economic issues, including workforce development and industry expansion and cultivation; natural and cultural resources; governance and civic institutions; environmental issues, including solid waste facilities; and other infrastructure systems, including repair, restoration, replacement, and improvement of public infrastructure such as water and wastewater treatment facilities, communications networks, and transportation infrastructure.

[24] See Pub. L. No. 114-187, 130 Stat. 549 (2016).

[25] See Pub. L. No. 115- 123, § 21210(e), 132 Stat. 64, 109 (2018).

provide financial oversight of these funds, Puerto Rico is developing an internal controls plan as well as management policies and procedures that will, in part, help provide financial monitoring. In the interim, FEMA has instituted a manual reimbursement process to mitigate risk and ensure fiscal accountability.

FEMA Obligated Almost $4 Billion, and Puerto Rico Expended Almost $1.7 Billion in Public Assistance Funding as of September 2018

As shown in figure 3, FEMA has obligated $3.63 billion (93 percent) for emergency work (categories A and B), and $151 million (4 percent) for permanent work (categories C through G) in Puerto Rico as of September 30, 2018.[26] An additional $136 million (3 percent) was obligated for management and administrative costs.

As of the end of fiscal year 2018, Puerto Rico expended about $1.7 billion (about 43 percent) of the almost $4 billion Public Assistance funds obligated by FEMA. Ninety-eight percent of this amount went toward emergency work projects in categories A and B. For example, the Puerto Rico Aqueduct and Sewer Authority expended almost $91 million to cover the costs of generator usage.

Aside from generators, one category B project by the Puerto Rico Emergency Management Agency repaired the emergency warning system for about $9.4 million. A third project put a temporary roof on a Puerto Rico Institute of Culture facility in Vieques for $4,000.

As shown in table 1, the majority of FEMA's obligations in Puerto Rico as of September 30, 2018, have been for emergency work categories because these projects began soon after the disasters struck and focused on

---

[26] FEMA data presented in this chapter on obligations and expenditures in Puerto Rico represent funding provided through FEMA's Public Assistance program. These data do not include obligations and expenditures for, among other things, direct federal mission assignments, in which a federal agency is tasked with providing eligible emergency work or debris removal services to a territory or state, or for other categories of mission assignments.

removing debris and providing assistance to address immediate threats to life and property.

In contrast, permanent work projects take time to identify, develop, and ultimately complete as they represent the longer-term repair and restoration of public infrastructure. Funds expended by Puerto Rico for permanent work have been mostly limited to roads and bridges (category C) because impassable roads like the one shown in figure 4 below impede the provision of critical services to citizens. They can also get in the way of other disaster recovery efforts.

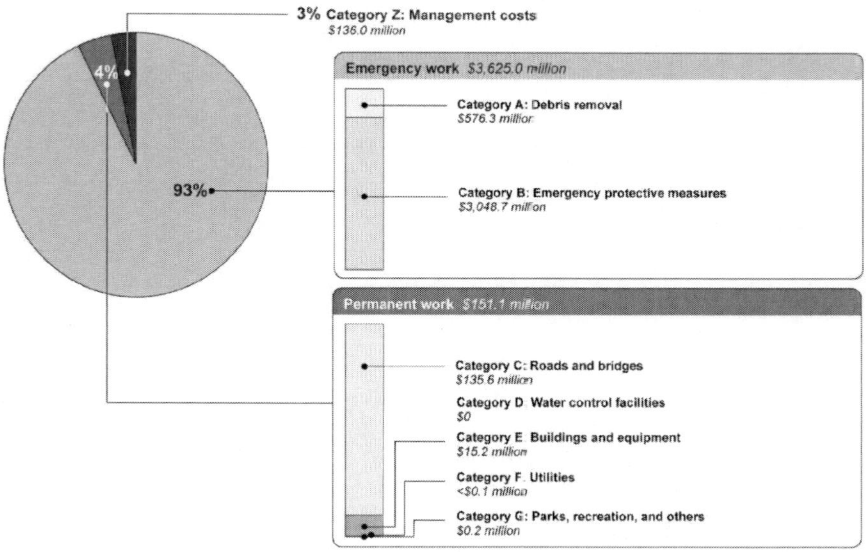

Source: FEMA's Emergency Management Mission Integrated Environment (EMMIE) and integrated Financial Management information system (IFMIS) data as of September 30, 2018. | GAO-19-256.

Figure 3. Federal Emergency Management Agency's Public Assistance Obligations to Puerto Rico by Category of Work, as of September 30, 2018.

Data Table for Figure 3. Federal Emergency Management Agency's Public Assistance Obligations to Puerto Rico by Category of Work, as of September 30, 2018

	Percentage	Dollars (in millions)
Emergency Work total	93%	$3,625.00
Emergency work Category A: Debris removal	--	$576.30
Emergency work Category B: Emergency protective measures	--	$3,048.70
Permanent work total	4%	$151.10
Permanent work Category C: Roads and bridges	--	$135.10
Permanent work Category D: Water control facilities	--	$0
Permanent work Category E: Buildings and equipment	--	$15.20
Permanent work Category F: Utilities	--	<$0.1
Permanent work Category G: Parks, recreation, and others	--	$0.20
Category Z: Management costs	3%	$136.00

Note: Numbers might not add up due to rounding. Emergency protective measures include activities to lessen the immediate threat to life, public health, or safety such as search and rescue operations and providing medical care and transport. Management costs are indirect costs, administrative expenses, and other expenses a recipient or subrecipient incurs in administering and managing the program that are not directly chargeable to a specific project. These data do not include obligations and expenditures for, among other things, direct federal mission assignments, in which a federal agency is tasked with providing eligible emergency work or debris removal services to a territory or state, or for other categories of mission assignments.

Expenditures for roads and bridges (category C) amount to approximately $32 million, while expenditures for other permanent work categories (DG) total approximately $1 million. According to FEMA officials, Public Assistance projects in categories D-G are still pending prioritization and formulation. For example, figure 5 below shows a recreational public space along the edge of a river in Maricao. The dashed line indicates where the iron railing and concrete paving used to continue, overlooking the river, before Hurricane Maria. As of September 2018, the municipality was awaiting FEMA assistance to begin restoration.

Table 1. Federal Emergency Management Agency's Public Assistance Obligations to Puerto Rico and Expenditures for Hurricanes Irma and Maria by Category of Work, as of September 30, 2018

Public Assistance Work Category	Obligated Amount (in millions)	Expended Amount (in millions)	Percent Expended
Debris Removal	$576.34	$137.50	24%
Emergency Protective Measures	$3,048.65	$1,513.76	50%
Roads and Bridges	$135.64	$32.40	24%
Water Control Facilities	—	—	—
Buildings and Equipment	$15.16	$1.24	8%
Utilities	$0.03	$0.01	29%
Parks, Recreational, and Other[a]	$0.24	$0.04	15%
Z: Management Costs[b]	$135.95	$1.49	1%
Total	$3,912.01	$1,686.43	43%

Legend: An em dash "—" indicates that zero funding corresponds with that category.

Source: FEMA's Emergency Management Mission Integrated Environment (EMMIE) and Integrated Financial Management Information System (IFMIS) data as of September 30, 2018 | GAO-19-256.

Note:

[a] Other refers to infrastructure such as playground equipment, mass transit facilities, piers, ports and harbors, and other items that do not fit in categories C-F.

[b] Management costs represent indirect costs, administrative expenses, or other expenses Puerto Rico or a subrecipient incurs in administering and managing the Public Assistance program that are not directly chargeable to a specific project. These data do not include obligations and expenditures for, among other things, direct federal mission assignments, in which a federal agency is tasked with providing eligible emergency work or debris removal services to a territory or state, or for other categories of mission assignments.

FEMA categorizes Puerto Rico's subrecipients as commonwealth public corporations, commonwealth agencies, municipalities, and all other entities. As shown in table 2 below, 89 percent of obligations, as of the end of September 2018, for Puerto Rico were awarded to commonwealth public corporations and commonwealth agencies, with 47 percent awarded to the Puerto Rico Electric Power Authority. Overall, about 43 percent of obligated funds have been expended.

Source: GAO. | GAO-19-256.

Figure 4. Collapsed Segment of Road Near Maricao, Puerto Rico, September 2018.

Source: GAO. | GAO-19-256.

Figure 5. Missing Rail in Public Space Along River's Edge, Maricao, Puerto Rico, September 2018.

Table 2. Federal Emergency Management Agency (FEMA)'s Public Assistance Obligations to Puerto Rico and Expenditures by Subrecipient (as categorized by FEMA) for Hurricanes Irma and Maria, as of September 30, 2018

Subrecipients	Percent of All Obligations to Puerto Rico	Obligated Amount (in millions)	Percent of Obligations Expended	Expended Amount (in millions)
All Commonwealth Public Corporations and Commonwealth Agencies	89%	$3,477.07	43%	$1,511.82
Puerto Rico Electric Power Authority (PREPA)	47%	$1,848.72	51%	$939.70
Puerto Rico Department of Housing	17%	$665.59	42%	$276.78
Puerto Rico Department of Transportation and Public Works	10%	$390.05	12%	$47.65
Central Recovery and Reconstruction Office (COR3)	3%	$107.01	1%	$1.49
Other Commonwealth Public Corporations and Other Commonwealth Agencies[a]	12%	$465.59	53%	$246.19
Municipalities of Puerto Rico	11%	$419.20	41%	$173.67
Other[b]	0%	$15.75	6%	$0.95
Total	100%	$3,912.01	43%	$1,686.43

Source: FEMA's Emergency Management Mission Integrated Environment (EMMIE) and Integrated Financial Management Information System (IFMIS) data as of September 30, 2018 | GAO-19-256.

Note: Numbers might not add up due to rounding. Subrecipient categories (i.e., Commonwealth Public Corporations, Commonwealth Agencies, Municipalities, Other) are as categorized by FEMA. Individual entities in this breakdown were selected on the basis that they comprised ten percent or more of the total Public Assistance obligations to Puerto Rico. Moreover, COR3 was added as well, due to its significant role in Puerto Rico's disaster recovery and Public Assistance fund administration.

[a] Other Commonwealth Public Corporations and Commonwealth Agencies as categorized by FEMA include entities such as the Puerto Rico Housing Finance Authority, the Puerto Rico Land Authority, the Puerto Rico Department of Education, the Puerto Rico National Guard, and many others.

[b] Other, as categorized by FEMA, include entities such as schools and universities, hospitals, community services, and nonprofit organizations. These data do not include obligations and expenditures for, among other things, direct federal mission assignments, in which a federal agency is tasked with providing eligible emergency work or debris removal services to a territory or state, or for other categories of mission assignments.

PUERTO RICO IS DEVELOPING AN INTERNAL CONTROLS PLAN AND RECOVERY MANAGEMENT POLICIES AND PROCEDURES TO PROVIDE OVERSIGHT OF FEDERAL RECOVERY FUNDS

As previously discussed, Puerto Rico designated COR3 to administer and manage the Public Assistance program in coordination with FEMA. As part of COR3's recovery oversight role, COR3 officials stated that they are developing an internal controls plan and recovery management policies and procedures with FEMA. According to COR3, these oversight documents will provide detailed guidance on grant application, procurement, payment and cash management, and financial monitoring and reporting, among other things. According to COR3 officials, they have held numerous meetings to coordinate with FEMA and have submitted drafts of the internal controls plan as well as management policies and procedures for FEMA's consideration. In addition, according to COR3 officials, COR3 plans to provide direct technical assistance related to federal grants management to Puerto Rico's cabinet-level agencies, public corporations, municipalities and other eligible subrecipients. As part of COR3's advisory role, COR3 is expected to help Puerto Rico's agencies, public corporations, municipalities, and some nonprofit entities formulate projects, draft cost estimates, and make funding requests, among other things.

FEMA Has Instituted an Interim Manual Reimbursement Process to Mitigate Risk

Federal grant award regulations allow FEMA to impose additional specific grant award conditions in specific circumstances, such as to mitigate risk and ensure fiscal accountability of the recipient or subrecipient.[27] According to FEMA, once FEMA obligates funds, the recipient is able to expend funds as necessary. However, in November 2017, according to FEMA officials, the agency instituted a manual reimbursement process for subrecipients in Puerto Rico for federal funds, including Public Assistance funds, to mitigate fiduciary risk and decrease the risk of misuse of funds. Specifically, FEMA officials stated that they decided to institute this process because the government of Puerto Rico had expended funds prior to submitting complete documentation of work performed.[28] According to FEMA officials, they also decided to institute the manual reimbursement process due to Puerto Rico's financial situation, weaknesses in internal controls, and the large amount of recovery funds, among other things.

This manual reimbursement process requires that COR3 fill out the Office of Management and Budget's Standard Form 270 and submit supporting documentation before obligated funds can be withdrawn by Puerto Rico through COR3 and reimbursed to subrecipients.[29] Subsequently, FEMA must review the submitted Standard Form 270 and all project documentation for completeness, compliance, and accuracy before disbursing funds to the recipient. In cases where FEMA requires additional documentation to process a Standard Form 270 request, FEMA will submit requests for information asking COR3 to supply the information needed for FEMA to complete the review.[30] FEMA officials said that they aim to complete the entire process described above within ten calendar days, or 15-20 calendar days if FEMA needs to request additional information from COR3. Additionally, FEMA officials stated that the manual reimbursement process is intended as a temporary

[27] See 2 C.F.R. §§ 200.205, 200.207; see generally 2 C.F.R. part 200. These additional award conditions may include requiring payments as reimbursements rather than advance payments; withholding authority to proceed to the next phase until receipt of evidence of acceptable performance within a given period of performance; requiring additional, more detailed financial reports, among other conditions.

[28] COR3 is responsible for collecting and submitting all documentation requested by FEMA from subrecipients.

[29] Standard Form 270: Request for Advance or Reimbursement.

[30] Additional documentation requested by FEMA to process a Standard Form 270 request includes, among other things, copies of contracts, cost of labor estimates, bid documents, and evidence of payment.

measure. They will cease the process once FEMA has reviewed the operational effectiveness of COR3's internal controls and approved the final internal controls plan, which are under review.[31]

FEMA, COR3, AND SUBRECIPIENTS REPORT INITIAL CHALLENGES WITH THE RECOVERY PROCESS

FEMA, COR3, and Puerto Rico municipal government officials from ten municipalities we interviewed reported initial challenges with the recovery process, including with Public Assistance alternative procedures. These concerns included (1) workforce capacity constraints, (2) a need for additional guidance, (3) delays related to choosing cost estimators, and (4) reimbursement for emergency work.[32]

Workforce Capacity Constraints

FEMA and municipality officials cited concerns about FEMA staff turnover and lack of knowledge about how the Public Assistance alternative procedures are to be applied in Puerto Rico.

---

[31] As part of our ongoing review, we will continue to monitor COR3's role in disaster recovery, the development of its recovery management policies, procedures, and plan, and FEMA's review of these efforts.

[32] As a part of our ongoing work on the alternative procedures process in Puerto Rico, we plan to conduct interviews with Puerto Rico government agencies in addition to municipalities who have been obligated funds from FEMA through the Public Assistance program. At the time of this chapter, we met with municipalities only. We chose to initially focus our work on municipalities to learn more about the disaster recovery process at the local government level, where officials are best positioned to speak directly to the impact of the Public Assistance program on the ground. According to the National Disaster Recovery Framework, local governments are primarily responsible for managing local recovery. Puerto Rico's 78 municipalities are the highest level of local government under the central-level government and each municipality has the legal authority over all matters of a municipal nature. Each municipal government has an executive branch, led by a municipal mayor, and a municipal legislature.

While several municipal officials we spoke to remarked positively on consistent communication with FEMA officials, municipal officials in six municipalities we visited cited high levels of turnover among FEMA staff as a challenge.[33] For example, officials in three municipalities said that discontinuity in FEMA personnel has caused them to have duplicative conversations with FEMA. An official from one municipality described the disruption that had been caused by repeated changes in FEMA personnel, especially when their point of contact at FEMA changed at least six times since the hurricanes.

FEMA officials acknowledged that more personnel with expertise in the alternative procedures process are needed to administer the Public Assistance program and assist subrecipients. According to FEMA officials, FEMA has leveraged existing expertise from personnel in the Federal Coordinating Officer Advisory Group to train new employees to increase workforce capacity. FEMA personnel from this group are rotating experts assigned to recovery issues to increase institutional understanding of alternative procedures and train local hires. According to FEMA officials, these local hires can serve as FEMA staff for up to one year before they become reservists.[34] In addition, FEMA officials stated that they have identified contractors with previous experience regarding alternative procedures to provide additional assistance to subrecipients.

Need for Additional Guidance

Municipal officials cited concerns about a lack of comprehensive guidance for the alternative procedures process. Specifically, officials in

---

[33] During interviews with officials from 10 selected municipalities, officials identified various challenges during the course of our discussions. When we report the number of municipalities that identified a particular challenge, this does not necessarily mean that the remaining municipalities did not also experience the challenge. It means that those municipalities did not raise the challenge during the course of our interviews.

[34] Reservists work on an intermittent basis and are deployed as needed. Reservists' activities can include interviewing disaster survivors; conducting and verifying damage assessments; providing administrative, financial, and logistical support; and performing a wide variety of other tasks as identified by staffing needs and operational requirements.

eight municipalities we interviewed cited problems with missing, incomplete, or conflicting guidance from FEMA.

In addition, officials in four municipalities stated that they are waiting on additional written instructions to establish more clear and consistent guidance. Officials from one municipality told us that the lack of written guidance has meant that the municipality has had to re-submit documents to FEMA multiple times to respond to changing guidance that they have received verbally. Additionally, four municipalities cited missing, incomplete, or conflicting guidance from COR3 as a challenge. However, one municipality noted that the quality of communication with COR3 has improved over time as COR3 has become more established.

According to FEMA officials, they are drafting supplemental guidance for the alternative procedures process with the goal of incorporating lessons learned from prior iterations of the alternative procedures. Similarly, according to COR3 officials, COR3 is currently developing additional guidance and standard operating procedures to help subrecipients, including municipalities and Puerto Rico government agencies, better understand FEMA Public Assistance grant requirements.

Delays Related to Choosing Cost Estimators for Puerto Rico

As mentioned previously, FEMA's guidance for alternative procedures requires that FEMA and Puerto Rico, through COR3, choose personnel with expertise in cost estimation to serve as a center of excellence, which will develop a cost estimating methodology. FEMA has chosen personnel to staff the center of excellence. However, in August 2018, FEMA officials told us that COR3 had not yet finalized their choice of personnel, which had delayed the cost estimation process.

Subsequently, COR3 officials told us that personnel have been identified to serve on the center of excellence and the final contracting process for these personnel is now in progress.

Reimbursement for Emergency Work

Officials in nine municipalities we spoke to said that they had not been fully reimbursed for emergency work they completed. Further, officials in five municipalities we interviewed stated that the lack of full reimbursement has caused financial hardships. For example, officials in three municipalities said that the lack of full reimbursement has meant that the municipalities have had to pause or delay recovery work due to lack of financial resources. A mayor in one municipality stated that they have scaled back some essential services, such as the frequency of garbage pick-up, while waiting for full reimbursement.

According to FEMA officials, delays in providing reimbursement were due to several factors including a loss in FEMA personnel to process reimbursement requests and a significant increase in the volume of reimbursement requests submitted by COR3 to FEMA. FEMA officials also stated an increasing need to make requests for information to COR3 due to a lack of documentation submitted at the time of the reimbursement request. In response to these factors, FEMA officials told us that they have undertaken new procedures with COR3. For example, according to FEMA officials, COR3 adopted procedures to review the completeness of documentation prior to submitting a reimbursement request to FEMA. FEMA officials stated that the agency is also holding weekly meetings with COR3 to increase coordination, and that FEMA increased the number of personnel devoted to reimbursement reviews. According to officials from FEMA and COR3, these steps have contributed to reduced delays.

Puerto Rico Developed an Economic and Disaster Recovery Plan in Response to the Bipartisan Budget Act of 2018 That Addresses Long- and Short-Term Needs

In response to the Bipartisan Budget Act, Puerto Rico submitted an economic and disaster recovery plan (recovery plan) to Congress on August 8, 2018.[35] The recovery plan defines the priorities, goals, and expected outcomes of Puerto Rico's recovery related to building government capacity for the recovery and strengthening of Puerto Rico's infrastructure, among other things. The recovery plan estimates infrastructure repair and recovery costs of $132 billion and total recovery costs of $139 billion for a time period starting in 2018 and ending in 2028. According to the recovery plan, COR3 will guide recovery investment and policy in the months and years ahead and is intended to serve as a focal point for strategic thought and management of Puerto Rico's recovery.[36]

The recovery plan is generally responsive to the directives outlined in the Bipartisan Budget Act.[37] For example, Puerto Rico submitted the plan to Congress within 180 days of enactment of the Bipartisan Budget Act. The recovery plan defines priorities, goals, and expected outcomes for Puerto Rico's recovery effort based on damage assessments conducted by sector. As mentioned earlier, Puerto Rico developed the recovery plan in coordination with FEMA and with support of the U.S. Department of Energy, the U.S. Department of Health and Human Services, and other

---

[35] See Pub. L. No. 115- 123, § 21210, 132 Stat. 64, 108-09 (2018).

[36] Central Office for Recovery, Reconstruction and Resiliency, Government of Puerto Rico, Transformation and Innovation in the Wake of Devastation: An Economic and Disaster Recovery Plan for Puerto Rico (August 8, 2018).

[37] Because section 21210 of the Bipartisan Budget Act includes directives with future implementation deadlines, we did not evaluate whether the plan is responsive to all directives related to the recovery plan. For example, the Act required Puerto Rico, in coordination with FEMA, to produce a report on the progress of achieving the goals set out in the recovery plan every 180 days after its release. The first such report is expected to be released in February 2019. Additionally, the Act required that FOMB use and incorporate damage assessments into its fiscal plan while the plan is in development and in effect. According to COR3, these damage assessments are not yet finalized.

federal agencies with responsibilities outlined in the National Disaster Recovery Framework.

Additionally, the FOMB of Puerto Rico certified the recovery plan on August 28, 2018, as directed in the Bipartisan Budget Act, but provided two caveats to its certification. First, FOMB expressed concern that the recovery plan lacks sufficient detail of funding sources and estimates a much greater amount of federal funding than the certified fiscal plan for Puerto Rico projects. The recovery plan states that at the time of its release, Puerto Rico had not undergone eligibility reviews in various federal funding programs, and therefore the ability to identify accurate funding sources was limited. COR3 officials confirmed that full recovery funding needs will not be known until all damage assessments are complete, and they will continue to identify and leverage all funding resources as they are made available. Second, FOMB indicated that the recovery plan does not address oversight of federal funds and the recovery process. While the Bipartisan Budget Act does not require specific mechanisms for oversight of federal funding as part of the recovery plan, according to COR3 officials, they plan to implement internal controls, policies, and procedures to provide oversight.

Puerto Rico's recovery plan outlines 276 "courses of action" (actions)— defined by the plan as "a collection of potential activities, policies, and other actions that could contribute to recovery"—selected by Puerto Rico to align with its future recovery vision. As shown in table 3 below, the actions reflect Puerto Rico's short-term and long-term recovery vision, organized into three areas. First, the recovery plan proposes "precursor" actions—those that serve as a foundation for all future actions—that will be prioritized for implementation. For example, the recovery plan includes actions to build capacity of municipalities to secure and manage recovery funds, and to improve the quality and volume of public data available to decision makers. Second, the recovery plan proposes a set of actions that aim to build the infrastructure and systems that support Puerto Rico's economy, society, and disaster resiliency, such as addressing vulnerabilities in Puerto Rico's electric grid. Finally, the recovery plan proposes a set of actions that address Puerto Rico's long-term recovery goals, such as developing and enhancing Puerto Rico's visitor economy.

Table 3. Categories of "Courses of Action" in Puerto Rico's Recovery Plan

"Precursor" Actions	Capital Investments	Long-Term Strategic Initiatives
Actions that aim to set the foundation for all other recovery projects and initiatives in seven areas: • Building government capacity to handle day-to-day business and recovery efforts • Availability of high-quality data to support decision making • Enhance local capacity to secure and manage recovery funds • Maximizing the impact of federal funds • Supporting local-level planning • Increasing ease of doing business in Puerto Rico • Reducing barriers to formal work and incentivizing workforce training	Actions to build a strong infrastructure and systems organized into nine areas: • Energy • Telecommunications • Water • Transportation • Housing • Public Buildings • Education • Health • Natural Environment	Actions to address Puerto Rico's long-term recovery goals in eight areas: • Ocean Economy • Visitor Economy • Emergency Services • Agricultural Modernization and Processing • Digital Transformation • 21st Century Workforce • Entrepreneurship • Advanced Manufacturing

Source: GAO analysis of Puerto Rico's recovery plan: Transformation and Innovation in the Wake of Devastation: An Economic and Disaster Recovery Plan for Puerto Rico, Central Office for Recovery, Reconstruction and Resiliency, Government of Puerto Rico (August 8, 2018). | GAO-19-256.

Most individual actions in the recovery plan include initial and recurring cost estimates for the time period from 2018 through 2028. The recovery plan describes all cost estimates as preliminary, and says that more specific cost estimates require completion of damage assessments and more details about the implementation of actions.[38]

To develop Puerto Rico's disaster recovery plan, FEMA assisted COR3 in retaining the Homeland Security Operational Analysis Center (HSOAC), a federally-funded research and development center operated by the RAND Corporation under contract with DHS. According to FEMA officials, FEMA provided funding and technical assistance, through contractor support, for Puerto Rico to develop the recovery plan, but COR3 and Puerto Rico will be responsible for its implementation.[39] These officials also stated that Puerto Rico received input and technical assistance from other federal departments, such as those in the Recovery Support Function Leadership Group led by FEMA.[40]

HSOAC developed the recovery plan in consultation with Puerto Rico by developing a preliminary sector-by-sector assessment of damages and needs caused by Hurricanes Irma and Maria across Puerto Rico. In conjunction with Puerto Rico's stated vision for the recovery process, HSOAC's damage assessment report provided the baseline needed to define, compare, and prioritize actions.

HSOAC worked in teams of sector-specific experts to develop and refine the actions by reviewing reports, proposals, best practices, and other literature. For example, in June 2018, the U.S. Department of Energy released a report on energy resilience for Puerto Rico's electric grid,

---

[38] According to the recovery plan, the methods and information used to develop cost estimates varied for each action, but wherever possible rough-order-of-magnitude cost estimates were developed to support high-level planning. Estimates include costs for repairing damaged assets to pre-hurricane conditions or improving them to meet environmental and engineering codes and standards, and when cost-effective, improving their disaster resiliency. Costs are associated with correcting pre-hurricane deficiencies, as well as addressing hurricane damage. Estimates include incremental operations and maintenance costs if those costs increased as a result of the hurricanes. Full operations and maintenance costs were used for facilities that were not being maintained prior to the hurricanes. Some actions, such as proposed policy changes, did not include cost estimates, though the plan acknowledged potential administrative costs. Each action includes a cost range, indicating that estimates will change depending on costs of equipment or labor, and on policy and implementation decisions.

[39] FEMA awarded HSOAC nearly $27.6 million to assist Puerto Rico in developing its recovery plan and any associated products.

[40] FEMA chairs the Recovery Support Function Leadership Group, an interagency body comprised of over 20 other federal agencies that coordinate operational, resource, and preparedness issues relating to interagency disaster recovery activities at the national level.

containing recommendations for Puerto Rico to consider when developing the Recovery Plan. HSOAC sought feedback from various subject matter experts and stakeholders while developing the recovery plan. According to FEMA officials, FEMA's joint recovery office delivered interim drafts of the recovery plan to federal agency partners and Puerto Rico for feedback. HSOAC also sought input from local-level stakeholders, including Puerto Rico's municipal governments. For example, HSOAC commissioned a survey of officials from municipalities to gauge the challenges they faced in the aftermath of the 2017 hurricanes. According to FEMA and HSOAC officials, the survey, along with other input provided by mayors led to the development of actions focused on building the capacity of municipal governments to support recovery efforts.

HSOAC officials noted that while the final recovery plan was submitted to Congress, they will continue to produce products that will assist Puerto Rico and their stakeholders in recovery implementation. HSOAC intends to release updated versions of the recovery plan, including updated damage and needs assessments. Other expected products include detailed descriptions and cost estimates for each action and a lessons learned report. The Bipartisan Budget Act also directs Puerto Rico to develop a public report on the progress made in achieving the recovery plan's goals every 180 days after submission.

FEMA officials explained that the recovery plan serves as a strategic, direction-setting plan for recovery, and does not provide step-by-step or site-by-site guidance on the recovery process. FEMA officials also acknowledged that there may be some overlap between some of the actions in the recovery plan and some of the permanent work funded through FEMA's Public Assistance program, but that it is COR3's responsibility to merge and coordinate such recovery efforts.

Agency Comments and Our Evaluation

We provided a draft of this chapter to DHS and the government of Puerto Rico for review and comment. In its comments, reproduced in appendix I, DHS summarized the amount of Public Assistance funding provided to Puerto Rico through fiscal year 2018. DHS also described FEMA's temporary manual reimbursement process instituted to mitigate risk and ensure fiscal accountability of taxpayer dollars, and stated that FEMA is committed to supporting Puerto Rico as it finalizes internal controls, management policies and procedures to oversee disaster recovery funds. DHS also provided technical comments, which we incorporated as appropriate.

The government of Puerto Rico provided comments that we reproduced in appendix II. In its comments, the government of Puerto Rico stated that in addition to what was discussed in this chapter, COR3 achieved progress and faced additional challenges.

This chapter is a part of an ongoing review of disaster recovery efforts in Puerto Rico. The remainder of our ongoing work will continue to examine Puerto Rico's recovery process, including implementation of the Public Assistance alternative procedures process and efforts by FEMA and Puerto Rico to oversee disaster recovery funds, including the manual reimbursement process.

Chris P. Currie
Director, Homeland Security and Justice Issues

LIST OF REQUESTERS

The Honorable Michael Enzi Chairman
Committee on the Budget
United States Senate

The Honorable Ron Johnson
Chairman

The Honorable Gary C. Peters
Ranking Member
Committee on Homeland Security and Governmental Affairs
United States Senate

The Honorable Rand Paul, MD
Chairman
Subcommittee on Federal Spending, Oversight and Emergency Management
Committee on Homeland Security and Governmental Affairs
United States Senate

The Honorable Maxine Waters Chairwoman
Committee on Financial Services
House of Representatives

The Honorable Sean Duffy
Ranking Member
Subcommittee on Housing, Community Development, and Insurance
Committee on Financial Services
House of Representatives

The Honorable Al Green
Chairman
Subcommittee on Oversight and Investigations
Committee on Financial Services
House of Representatives

The Honorable Bennie Thompson
Chairman
Committee on Homeland Security
House of Representatives

The Honorable Peter T. King
Ranking Member
Subcommittee on Emergency Preparedness, Response, and Recovery
Committee on Homeland Security
House of Representatives

The Honorable Elijah Cummings
Chairman
Committee on Oversight and Reform
House of Representatives

The Honorable Nydia Velázquez
Chairwoman
Committee on Small Business
House of Representatives

The Honorable Peter DeFazio
Chairman

The Honorable Sam Graves
Ranking Member
Committee on Transportation and Infrastructure
House of Representatives

The Honorable Emanuel Cleaver, II
House of Representatives

The Honorable Michael McCaul
House of Representatives

The Honorable Gary Palmer
House of Representatives

The Honorable Ann Wagner
House of Representatives

APPENDIX I: COMMENTS FROM THE DEPARTMENT OF HOMELAND SECURITY

U.S. Department of Homeland Security
Washington, DC 20528

Homeland Security

March 4, 2019

Chris P. Currie
Director, Homeland Security and Justice Issues
U.S. Government Accountability Office
441 G Street, NW
Washington, DC 20548

Re: Management Response to Draft Report GAO-19-256, "Puerto Rico Hurricanes: Status of FEMA Funding, Oversight, and Recovery Challenges"

Dear Mr. Currie:

Thank you for the opportunity to review and comment on this draft report. The U.S. Department of Homeland Security (DHS) appreciates the U.S. Government Accountability Office's (GAO) work in planning and conducting its review and issuing this report.

We are pleased that GAO's draft report recognizes (1) the extent of damages the 2017 hurricanes caused throughout Puerto Rico, (2) that the Federal Emergency Management Agency (FEMA) has obligated nearly $4 billion through fiscal year 2018 in Public Assistance funding for Puerto Rico emergency work projects, as well as the repair of its public infrastructure, among other things, (3) both FEMA and Puerto Rico have taken actions to provide oversight of these Federal recovery funds, and (4) challenges remain for FEMA and Puerto Rico's Central Office for Recovery, Reconstruction and Resilience (COR3) to fulfill their respective responsibilities to rebuild damaged infrastructure and restore critically-needed services, while ensuring proper stewardship of taxpayer dollars.

One way FEMA is mitigating risk and ensuring fiscal accountability of taxpayer dollars is through a manual reimbursement process prohibiting Puerto Rico's unrestricted withdrawal of Public Assistance funds. This temporary process was instituted due to Puerto Rico's financial situation, weaknesses in internal controls, and the large scope of recovery funds, among other things. The process helps ensure that funds are expended only for projects that have appropriate supporting documentation, thus reducing the chance that FEMA will later need to recover funding found to be associated with unsupported costs. As the draft report relates and to its credit, Puerto Rico is developing an internal controls plan as well as management policies and procedures that will, in part, help provide financial monitoring. This manual process will end once FEMA has reviewed the operational effectiveness of COR3's internal controls and approved the final internal controls plan, which are currently under review.

FEMA remains committed to supporting COR3 efforts to finalize guidance on grant application, procurement, payment and cash management, and financial monitoring and reporting for both COR3 and its sub recipients. DHS remains committed to supporting our citizens and first responders before, during and after disasters.

We noted that the draft report did not contain any recommendations. Technical comments were previously provided under separate cover. We look forward to GAO's continued work related to Puerto Rico hurricane recovery efforts and any formal recommendations once that work is complete.

Again, thank you for the opportunity to review and comment on this draft report. Please contact me if you have any questions. We look forward to working with you in the future.

Sincerely,

JIM H. CRUMPACKER, CIA, CFE
Director
Departmental GAO-OIG Liaison Office

APPENDIX II: COMMENTS FROM THE GOVERNMENT OF PUERTO RICO

GOVERNMENT OF PUERTO RICO
Central Recovery and Reconstruction Office of Puerto Rico

January 30, 2019

The Honorable Gene L. Dodaro
Comptroller General of the United States
U.S. General Accountability Office
441 G Street, NW
Washington, DC 20548

Re: COR3 comments to February 2019 GAO Report: Puerto Rico Hurricanes: Status of FEMA Funding, Oversight and Recovery Challenges, GAO-19-256

Dear Mr. Dodaro,

I am writing with regard to the U.S. Government Accountability Office (GAO) report, GAO-19-256, *Puerto Rico Hurricanes: Status of FEMA Funding, Oversight, and Recovery Challenges*. The Government of Puerto Rico and the Central Recovery and Reconstruction Office (COR3) would like to thank GAO's engagement team for their efforts to date in compiling this document and for the opportunity to review and provide comment for inclusion in the final version of the GAO Report. COR3 believes that there are areas within the document where progress by COR3 has been achieved and can be further defined. Additionally, though some were described in the report, there are additional challenges to Puerto Rico's recovery that remain and need to be addressed. There exists a strong partnership between COR3 and local FEMA leadership and we are confident that our continued collaboration and mutual desire to support the U.S. citizens of Puerto Rico will ensure success in our recovery efforts.

Below are the official comments and feedback of the Government of Puerto Rico.

1. **COR3 Committed to Transparency and Accountability in Recovery:** COR3 has established a comprehensive and effective system, which includes policies and procedures, to manage federal funding and promote accountability and transparency. This system is modeled after systems successfully implemented by other states in disaster recovery to facilitate a strong recovery for Puerto Rico. COR3 has implemented policies and procedures to support the strategic focus of short and long-term recovery efforts to include appropriate internal controls, procurement processes, grant monitoring, cash and property management processes, and reporting. Additionally, COR3 is developing a strong subrecipient monitoring, training, and technical assistance program to support the recovery of other Puerto Rico government agencies, municipal governments, and private non-profits.

2. **Need Consistent FEMA Guidance on Implementation of Section 428:** FEMA has required Puerto Rico to develop all permanent work processes using the provisions of Section 428 of the Stafford Act. Section 428, also known as alternative procedures, has never been applied on such a large scale before and the lack of experienced FEMA staff or clear and consistent guidance from FEMA has contributed to untenable delays in recovery efforts. Additionally, the difficulty with Section 428 is that it requires a "fixed cost estimate," which, once agreed upon by FEMA, the recipient, and the subrecipient, cannot be changed. Under a typical PA program project, the grant amount is estimated and if circumstances are discovered during the repair/replacement period that change the scope of work or cost, the project can be amended or versioned. The fixed cost estimate requirement, the review of a project before obligation of funding is onerous and time consuming, which is causing the entire recovery process to be delayed.

3. **Extension of 428 Timeline Needed to Address Delays in Fixed Cost Estimate Approval by FEMA:** FEMA has set a deadline of October 2019 for the completion of all Fixed Cost Estimates for Permanent Work. As of January 2019, only eleven Fixed Cost Estimates have been completed and agreed upon. Due to the onerous and disadvantageous nature of this process, COR3 does not believe that it is possible for FEMA and COR3 to meet the October 2019 deadline and an extension will be required.

4. **Need to Remove Unnecessary Duplication in 428 Implementation:** Currently, both FEMA and the subrecipients are doing separate and detailed analysis of damages (Damage Description and Dimensions (DDD)) and costs. This process is duplicative and time-consuming and has contributed to the delay in obligation of the PA projects. Both the Sandy Recovery Act of 2013 and the Disaster Recovery Reform Act of 2018, encourage FEMA to rely on the certified cost estimate of the subrecipient's licensed engineer. COR3 has recommended to FEMA that they follow the Congressional intent of these two pieces of legislation concerning the damage assessment process, and rely on the product of a professionally licensed engineer, thus reducing duplication of effort and expediting the process of funding eligible projects.

5. **COR3 Ready for FEMA to End 270 Manual Drawdown Process That is Causing Undue Burdens and Delays in Recovery:** FEMA implemented a manual drawdown process, more commonly known as the "270 process," for reimbursements under the PA program. The process places an undue burden on the applicant and subrecipients that significantly delays the reimbursement process. Because of the economic condition of most subrecipients across Puerto Rico, a lengthy delay in reimbursement means they may not be able to pay contractors in a timely manner. Under the relevant provisions of Titles 2 and 44 of the Code of Federal Regulation (CFR), states/recipients are in fact responsible for managing the reimbursement of funding to subrecipients and have the flexibility under the CFR provision to develop a system that supports the needs of subrecipients. It is our belief that FEMA has never implemented such manual drawdown for any previous disaster in the nation. However, following guidance from FEMA, COR3 sourced highly qualified subject matter specialists experienced with the FEMA programs. COR3 developed compliant policies and procedures (patterned after those implemented in other states), which have been reviewed and fully coordinated with FEMA. COR3 has demonstrated to FEMA that we are prepared to assume the process; however, FEMA has placed additional burdensome conditions on COR3 that will not allow us to be responsive to the needs of subrecipients. We understand and are willing to assume the full responsibility and risk associated with this role and view this as necessary to facilitate the speed of the recovery. Without the flexibility to manage the process as allowed by the CFR, COR3 cannot accept management of the reimbursement process.

6. **Congressional Intent on Pre-Existing Conditions Waiver Must be Respected for Successful Recovery:** On February 9, 2018, President Donald Trump signed into law the Bipartisan Budget Act of 2018 (BBA), P.L. 115-123. The legislation passed by comfortable margins in both the U.S. Senate (71-28) and the U.S. House of Representatives (240-186). The legislation extended funding of federal government operations, increased the allowable discretionary spending levels for defense and non-defense programs, lifted the debt ceiling for borrowing by the federal government, and provided for disaster relief funding. The legislation included several provisions that impact Puerto Rico. One provision of the BBA, Section 20601, gives FEMA the authority to provide assistance to restore disaster-damaged facilities or systems to industry standard and to restore functionality of the disaster-damaged facility or system without regard to pre-disaster condition. This provision was included in the legislation and approved by Congress and the Administration for one very important reason: to ensure that the large investment of federal taxpayer monies into the recovery efforts on Puerto Rico and the U.S. Virgin Islands did not end up throwing "good money after bad" by restoring facilities to a condition that would only cause them to continue to deteriorate, put people back into at-risk situations in sub-standard facilities, and result in both buildings and infrastructure not being reconstructed in a manner that meets current best-practices and standards. To do otherwise would have been shortsighted and would not only waste the many billions of dollars of taxpayer funds that would be invested in Puerto Rico and the U.S. Virgin Islands, but also put the taxpayer at extraordinary risk the next time a hurricane or other event caused widespread impacts to the islands. Unfortunately, since the passage of the BBA, the Administration has taken many steps to narrow or eliminate eligibility under Section 20601. One of our main concerns is FEMA's recent position that it will not implement BBA Section 20601 and that Section 406 of the Stafford Act (PA Program) regulations trump the BBA is counter Congressional intent to provide Puerto Rico sufficient funding to allow structures to be rebuilt without funding reductions for pre-existing condition that they cannot afford, and without missing opportunities to improve, modernize, and make their facilities and systems more resilient from future disasters. We believe these actions, taken in total, represent an effort to undermine the Congressional intent of the BBA Section 20601, and it also puts Puerto Rico at great risk that its entire recovery effort could "stall" due to lack of necessary funding to restore damaged buildings and infrastructure.

In closing, I want to thank GAO for its efforts in preparing this report. I hope that the comments contained herein are helpful. Should you have any questions or would like to discuss our comments further, please feel free to contact me at your convenience.

Sincerely,

Omar J. Marrero, Esq.
Executive Director
Governor's Authorized Representative

cc: Chris P. Currie. Director, Homeland Security and Justice Issues, GAO

In: Hurricanes, Wildfires and Flooding
Editor: Lydie Yohan
ISBN: 978-1-53616-362-9

Chapter 4

MAJOR DISASTER ASSISTANCE FROM THE DRF: PUERTO RICO*

Bruce R. Lindsay

ABSTRACT

The Disaster Relief Fund (DRF) is managed by the Federal Emergency Management Agency (FEMA) and is the primary source of federal assistance following a major disaster declaration. The authority to declare a major disaster is provided to the President under the Robert T. Stafford Emergency Relief and Disaster Assistance Act (P.L. 93-288, as amended; Stafford Act).

The information below includes only DRF funding provided to Puerto Rico as a result of assistance associated with a major disaster declaration declared between FY2000 and FY2016. This information does not include other assistance provided through the DRF or disaster assistance provided by other federal sources, such as Small Business Administration disaster loans. In FY2017, two major disasters were declared for Puerto Rico. Those major disaster declarations are not included in the analysis below.

* This is an edited, reformatted and augmented version of Congressional Research Service, Publication No. IF10755, dated October 16, 2017.

FEMA administers disaster relief through regional offices around the country. Puerto Rico is part of Region II, headquartered in New York, NY. Puerto Rico is considered a state under Section 102 of the Stafford Act (42 U.S.C. 5122) for the purposes of that act.

MAJOR DISASTER DECLARATIONS: FY2000-FY2016

A total of 11 major disaster declarations were made for Puerto Rico between FY2000 and FY2016. During that time, there were no requests for major disaster declarations that were denied. The approved declarations led to $833 million in actual and projected federal obligations from the DRF. Actual and projected obligations are used throughout this In Focus to account for the total amount of federal assistance ultimately obligated. This funding does not include the assistance that was provided directly by the territory, either as a cost-share for federal assistance or through the territory's own authorities and programs.

There was a high level of variation in the amount of actual and projected funding obligated for major disasters each year, with more than $511 million in actual and projected obligations for disasters in FY2004 alone. Figure 1 displays the actual and projected obligations to Puerto Rico each fiscal year. The highest number of declarations in a single year was three, which occurred in FY2011.

Most Costly Declarations

The declaration associated with the largest actual and projected obligations from the DRF to Puerto Rico was for Tropical Storm Jeanne in FY2004. This declaration resulted in more than $431 million in actual and projected obligations to Puerto Rico from the DRF.

The next two most costly declarations were for Hurricane Irene in FY2011 and severe storms and flooding in FY2009.

Local Impacts

From FY2000 to FY2016, all 78 local and territorial governments in Puerto Rico received aid under major disaster declarations. The municipality of Utuado received aid for the highest number of declarations during this period, with nine.

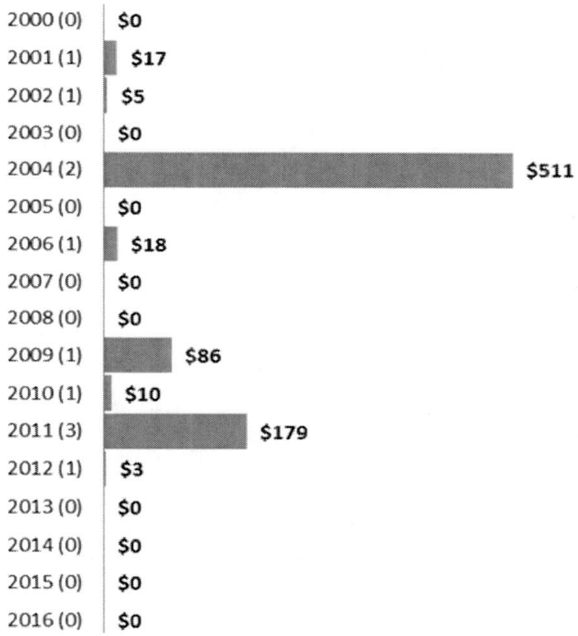

Source: CRS analysis of FEMA DRF obligation data as of August 2017.

Notes: Actual and projected obligations are recorded in the fiscal year in which the major disaster was declared. Only obligations from the FEMA DRF are included.

Figure 1. DRF Actual and Projected Obligations for Declared Major Disasters in Puerto Rico, by Fiscal Year. Figures are in millions of dollars, truncated. The number of declarations per year is in parentheses.

TYPES OF ASSISTANCE PROVIDED FOR MAJOR DISASTERS

A major disaster declaration can authorize funding for different purposes, depending on the needs of the territory. These purposes include:

- *Public Assistance*, which is used by territorial or local governments, or certain private nonprofit organizations to provide emergency protective services, conduct debris removal operations, and repair or replace damaged public infrastructure;
- *Individual Assistance,* which provides direct aid to impacted households;
- *Hazard Mitigation Assistance*, which funds mitigation and resiliency projects and programs, typically across the entire territory;
- *FEMA administrative costs* associated with each disaster declaration; and
- *Mission Assignments*, which tasks and reimburses other federal entities that provide direct disaster assistance.

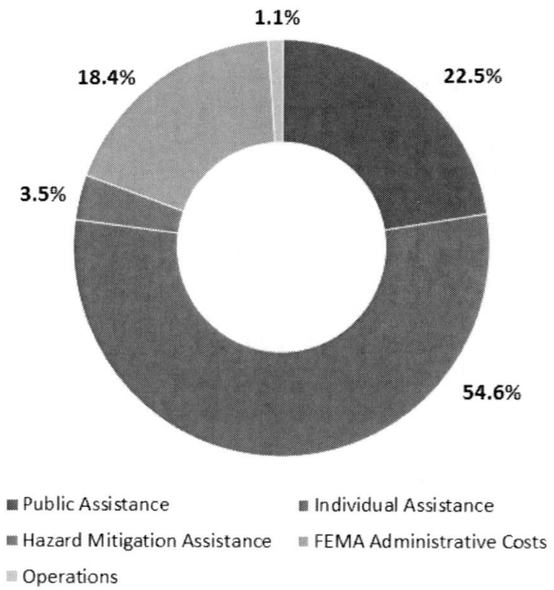

Source: CRS analysis of FEMA DRF obligation data as of August 2017.

Note: The specific assistance within each of the identified types may have changed in the period FY2000 to FY2016. "Operations" is a category provided by FEMA that includes three types of Mission Assignments and Urban Search and Rescue.

Figure 2. DRF Assistance for Puerto Rico, by Type (FY2000-FY2016).

The decision concerning which types of assistance to provide is made either when the major disaster is declared or when the declaration is amended. For many major disasters, all of the assistance types outlined above are authorized. For others, some assistance types are not authorized. Figure 2 compares the actual and projected obligations for different types of assistance provided to Puerto Rico as a result of a major disaster declaration from FY2000 to FY2016.

OTHER FEMA ASSISTANCE FUNDED THROUGH THE DRF

In addition to the major disaster assistance described above, there are other forms of assistance that are funded through the DRF. The funding associated with these types of assistance typically results in lower obligation levels than assistance provided as a result of major disaster declarations, although there is significant variation across incidents.

- *Emergency Declarations* are often made at the time a threat is recognized in order to assist territorial and local government efforts prior to an incident.
- *Fire Management Assistance Grants (FMAGs)* provide aid for the control, management, and mitigation of fires.

FEMA ASSISTANCE NOT FUNDED THROUGH THE DRF: NATIONAL FLOOD INSURANCE PROGRAM

Floods represent a majority of all major disaster declarations nationwide. One of the primary sources of assistance for flooding events is the National Flood Insurance Program (NFIP), which is not funded through the DRF.

In: Hurricanes, Wildfires and Flooding ISBN: 978-1-53616-362-9
Editor: Lydie Yohan

Chapter 5

PUERTO RICO—STATUS OF ELECTRIC POWER RECOVERY*

Richard J. Campbell

ABSTRACT

On September 20, 2017, Hurricane Maria made landfall in Puerto Rico as a Category 4 storm with sustained wind speeds of over 155 miles per hour. At that time, the Commonwealth of Puerto Rico was already in recovery mode following the glancing blow struck by Hurricane Irma on September 6, 2017, which left 70% of electricity customers without power. Puerto Rico's office of emergency management reported that Hurricane Maria had incapacitated the central electric power system, leaving the entire island without power as the island's grid was essentially destroyed. Even before the 2017 hurricane season, Puerto Rico's electric power infrastructure was known to be in poor condition, due largely to underinvestment and the perceived poor maintenance practices of the Puerto Rico Electric Power Authority (PREPA).

The primary focus of territorial and federal efforts thus far has largely been on restoring electric power in Puerto Rico. The new

* This is an edited, reformatted and augmented version of Congressional Research Service Publication No. R45263, dated Updated July 10, 2018.

hurricane season in the Atlantic Basin (comprised of the Atlantic Ocean, the Caribbean Sea, and the Gulf of Mexico) began on June 1, 2018, and will last until November 30.

A large part of the Federal Emergency Management Agency's (FEMA's) role in Puerto Rico was centered on coordinating the restoration of electric power. For this task, FEMA made available about 900 portable generators (including several generation units providing over 50 Megawatts of capacity at the partially operating Palo Seco station in San Juan), and contracted with the U.S. Army Corps of Engineers (USACE) to oversee restoration of the electric transmission system, which was damaged extensively. PREPA itself has largely been focused on the restoration of the electric distribution system, and service connections to its customers. In Puerto Rico, USACE is not only restoring emergency power but also leading initial grid repair. With the end of the grid repair part of USACE's mission assignment on May 18, 2018, almost 99% of PREPA's customers have seen their electric service restored.

Section 21101 of the Bipartisan Budget Act of 2018 (P.L. 115-123) provided $28 billion in supplemental appropriations to the U.S. Department of Housing and Urban Development (HUD), under the Community Development Block Grant (CDBG) fund. Of this amount, up to $2 billion was made available (until expended) for "enhanced or improved electrical power systems."

Approximately $2 billion in CDBG funds were made available to "enhance or improve" any electric power systems damaged by Hurricane Maria. To help guide the process of rebuilding Puerto Rico's grid, the U.S. Department of Energy (DOE) reports that five national laboratories have collaboratively built a model of Puerto Rico's electricity system to test how to place microgrids, determine where to place power lines underground, and test siting of renewable energy projects where they can be sheltered from damage by extreme weather events. DOE believes its modeling efforts can therefore help guide HUD and FEMA CDBG investments to improve the power grid in Puerto Rico. DOE says that it plans to complete a "resilient grid model" to prioritize investments for "transmission, distribution, new generation, energy storage, microgrids, and strategic power reserves." This would allow potential impacts on other critical infrastructure such as the petroleum, natural gas, and telecommunications sectors to be estimated.

With the power restoration effort almost finished, the next focus for authorities will likely be on rebuilding Puerto Rico's electricity transmission and distribution systems, as a modern system built to U.S. industry standards is more likely to survive extreme weather events. Once the backbone of the infrastructure is in place, then the major effort of making the electricity system of Puerto Rico more resilient can follow in earnest. Building resilience would likely require improvements that go beyond even modernization and rebuilding. Resilience may require additional improvements to the system aimed at better withstanding the effects of extreme weather events.

BACKGROUND

On September 20, 2017, Hurricane Maria made landfall in Puerto Rico as a Category 4 storm with sustained wind speeds of over 155 miles per hour. At that time, the Commonwealth of Puerto Rico (the Commonwealth) was already in recovery mode following the glancing blow struck by Hurricane Irma on September 6, 2017, which left 70% of electricity customers without power. Puerto Rico's office of emergency management reported that Hurricane Maria had incapacitated the central electric power system, leaving the entire island without power as the island's grid was essentially destroyed. After Maria, officials were estimating that many of Puerto Rico's 3.5 million people could be without electricity for up to six months.

Even before the 2017 hurricane season, Puerto Rico's electric power infrastructure was known to be in poor condition, due largely to underinvestment and the perceived deficient maintenance practices of the Puerto Rico Electric Power Authority (PREPA). With the poor state of the electricity system (physically, organizationally, and financially),[1] and a

[1] In his written statement, Governor Ricardo Rosselló, governor of Puerto Rico, stated "On the eve of Hurricane Maria, Puerto Rico was literally bankrupt—having sought Title III protection for the Commonwealth and certain of its instrumentalities, including the Puerto Rico Electric Power Authority (PREPA), due to a mountain of over $80 billion of debt that it could no longer service with available revenues." Also, from the written statement of Ricardo Ramos, Executive Director of PREPA, "When I became Executive Director of the Puerto Rico Electric Power Authority in March 2017, I knew that PREPA faced significant challenges including a declining revenue base, a shrinking workforce, a backlog of unmet or deferred maintenance, key major equipment beyond its useful life, and of course, the many challenges associated with Puerto Rico's and PREPA's financial condition." U.S. Congress, Senate Committee on Energy and Natural Resources, *Written Statement of Hon. Ricardo Rosselló, Governor of Puerto Rico*, Hearing on Hurricane Recovery Efforts in Puerto Rico and the U.S. Virgin Islands, 115th Cong., 1st sess., November 14, 2017, p. 2. U.S. Congress, Senate Committee on Energy and Natural Resources, *Written Statement of Ricardo Ramos, Executive Director of PREPA*, Hearing on Hurricane Recovery Efforts in Puerto Rico and the U.S. Virgin Islands, 115th Cong., 1st sess., November 14, 2017, p. 1.

perceived lack of transparency with regard to decisions (both before and since Hurricane Maria),[2] discussions have already started about how the electricity system in Puerto Rico would be rebuilt, and under what regulatory regime it will operate.

The primary focus thus far has largely been on restoring electric power in Puerto Rico, while the task of rebuilding the grid in Puerto Rico to modern standards is expected to follow. The new hurricane season in the Atlantic Basin (comprised of the Atlantic Ocean, the Caribbean Sea, and the Gulf of Mexico) began June 1, 2018, and will last until November 30.

Congress continues to follow the recovery of Puerto Rico from the 2017 hurricanes, and the restoration of power. The electric power infrastructure for transmission and delivery of power was largely destroyed by the hurricanes, and PREPA's aging power generation facilities have been struggling to provide electricity. Whether, and how, Puerto Rico rebuilds its system into a reliable, efficient, and resilient electricity system will be of key interest to Congress.

STATUS OF ELECTRIC POWER RESTORATION

A large part of the Federal Emergency Management Agency's (FEMA's) role in Puerto Rico was centered on coordinating the restoration of electric power. For this task, FEMA brought about 900 portable generators (including several generation units providing over 50 Megawatts (MW) of capacity at the partially operating Palo Seco electric power generation station in San Juan), and tasked the U.S. Army Corps of Engineers (USACE) to oversee restoration of the electric transmission system, which was damaged extensively. PREPA itself has largely been focused on the restoration of the electric distribution system, and service

[2] Ross Marchand, "To Get the Lights Back On, Puerto Rico Must Boost Transparency," *The Hill*, October 26, 2017, http://thehill.com/opinion/finance/357326-to-get-lights-back-on-puerto-rico-must-boost-transparency.

connections to its customers.[3] Both USACE and PREPA have used contractors to do much of the transmission and distribution system repairs.[4] With the end of the grid repair part of USACE's mission assignment on May 18, 2018, almost 99% of PREPA's customers have seen their electric service restored. However, over 16,000 customers (in mostly rural parts of the island) were still without electricity after eight months.

The power restoration mission has been described as incomplete by the USACE, with much of the work to date essentially to patch up the electric system to bring power back to the people of Puerto Rico.[5] FEMA has agreed to leave over 700 portable generators on Puerto Rico past the end of the USACE's grid repair mission.[6]

FEMA and Grid Restoration Assistance[7]

PREPA is a public power utility[8] owned by the Commonwealth of Puerto Rico, and is the largest supplier of electricity in the Commonwealth.

---

[3] PREPA's power transmission system consists of 1,100 miles of 115 kilovolt (kV) and 230 kV transmission lines. In addition, there are approximately 1,375 miles of 38kV subtransmission level lines including 55 miles of undersea lines that connect the main island with the islands of Vieques and Culebra. The hurricanes knocked out of service over 2,400 miles of transmission lines (including 1,375 miles of subtransmission lines operating at 38 kV), and 30,000 miles of distribution lines, representing almost the entire transmission and distribution infrastructure of the island. See CRS Report R45023, *Repair or Rebuild: Options for Electric Power in Puerto Rico*, by Richard J. Campbell, Corrie E. Clark, and D Andrew Austin.

[4] Andrew Scurria, "Puerto Rico Contract Spending Accelerates with Army Corps Exit," *Wall Street Journal*, March 2, 2018, https://www.wsj.com/articles/puerto-rico-contract-spending-accelerates-with-army-corps-exit-1519948452.

[5] Voice of America, *Army Corps Quits Puerto Rico Saying Job is Incomplete*, May 18, 2018, https://www.voanews.com/a/army-corps-leaving-puerto-ricans-without-power/4399485.html.

[6] Carlos Giusti and Michael Weissenstein, "Corps Leaving Puerto Rico with Hurricane Recovery Unfinished," Associated Press, May 18, 2018, https://apnews.com/cba884b0d5664368a18dfe7ea1b18e75/Corps-leaving-PuertoRico-with-hurricane-recovery-unfinished.

[7] See CRS Report R45023, *Repair or Rebuild: Options for Electric Power in Puerto Rico*, by Richard J. Campbell, Corrie E. Clark, and D. Andrew Austin.

[8] "Public power utilities are operated by local governments to provide communities with reliable, responsive, not-forprofit electric service. Public power utilities are directly accountable to

As a public utility, PREPA is an eligible applicant that can receive federal assistance through the Federal Emergency Management Agency (FEMA). In particular, FEMA provides grant assistance through the Public Assistance Grant Program (PA Program) for the repair, restoration, and replacement of public facilities, as defined by law, in states and communities that have received major or emergency disaster declarations through the Robert T. Stafford Disaster Relief and Emergency Assistance Act (Stafford Act, P.L. 93-288, as amended).[9]

Role of the Army Corps of Engineers

USACE supports the Department of Homeland Security in carrying out the National Response Framework. If requested by the President and the affected governors under the Stafford Act, USACE's primary role under the framework is to provide emergency support in areas of public works and engineering.[10] These USACE activities are funded through the Disaster Relief Fund and not through direct appropriations to the agency. In Puerto Rico, USACE not only restored emergency power but also led initial grid repair. USACE leadership in grid repair as part of domestic disaster recovery is a novel development.[11]

On September 30, 2017, the FEMA Administrator also tasked USACE to work with PREPA, Department of Energy (DOE), and FEMA to provide a unified effort to repair Puerto Rico's power grid. USACE was tasked with leading the planning, coordination, and integration of the electric

---

the people they serve through local elected or appointed officials." American Public Power Association, *About Public Power*, 2017, http://publicpower.org/.

[9] For an overview of the PA Program, see CRS Report R43990, *FEMA's Public Assistance Grant Program: Background and Considerations for Congress*, by Jared T. Brown and Daniel J. Richardson.

[10] Other tasks typical for USACE include technical assistance; engineering; construction management; emergency contracting; and repair of power systems, public water and wastewater systems, and solid waste facilities.

[11] Each disaster often results in FEMA adjusting assignments to meet the needs and demands of the specific situation. Furthermore, USACE is the entity designated with emergency contracting under the National Response Framework (NRF). USACE was also active in assisting PREPA with emergency dam-safety measures at its Guajataca dam. USACE actions to reopen navigation through channel assessments and clearing has allowed for improved delivery of materials, equipment, and personnel to the island for grid repair and other recovery activities.

power grid repair. The Unified Command Group (comprised of the Corps, FEMA, PREPA, and Puerto Rico's Restoration Coordinator) was organized to evaluate and coordinate efforts for the work of restoring power to prestorm electricity customers (i.e., not new customers, poststorm).[12]

PREPA's electric power generation plant at Palo Seco in San Juan is not operating at full power due to the condition of the facility.[13] FEMA installed two "mega" 30 MW mobile diesel-fueled generators to stabilize power generation in the heavily populated region around San Juan.[14] FEMA also installed several smaller mobile generators at Palo Seco to provide additional power at times of peak electric demand. Another large mobile 30 MW generator is installed at the Yabucoa Power Plant to help stabilize power generation in the southeastern region of Puerto Rico. USACE personnel are assisting with the operation and maintenance of the mobile generators.

FEMA FUNDING FOR 2017 DISASTERS[15]

Supplemental Appropriations Bills

In light of the hurricanes and other disasters in 2017, Congress passed three supplemental appropriations bills in response to Administration requests made in September, October, and November 2017. Amounts for electric grid restoration in Puerto Rico were not specified, even though the third supplemental does allow for $2 billion in funding for "enhanced or

---

[12] U.S. Congress, Senate Committee on Energy and Natural Resources, Statement of Charles Alexander. Jr., Army Corps of Engineers, Update on the Restoration of Puerto Rico's Electric Infrastructure, 115th Cong., 2nd sess., May 8, 2018, pp. 3-4.

[13] Javier Colón Dávila, "Engineers to Assess Whether Palo Seco Can Operate," ElNuevodia, October 20, 2017, https://www.elnuevodia.com/english/english/nota/engineerstoassesswhetherpalosecocanoperate-2367636/.

[14] Approximately 70% of Puerto Rico's power generation is in the southern region of Puerto Rico, while approximately 70% of power demand is in the northern regions.

[15] See CRS Report R45084, 2017 Disaster Supplemental Appropriations: Overview, by William L. Painter.

improved electrical power systems" for all areas affected by Hurricane Maria.

First Supplemental, September 2017 (P.L. 115-56, Division B)

On September 1, 2017, the Trump Administration requested $7.85 billion in supplemental funding. On September 6, the House passed the relief package requested by the Administration as an amendment to H.R. 601. On September 7, the Senate passed the bill further amended to include an additional $7.4 billion for disaster relief through the Department of Housing and Urban Development's (HUD's) Community Development Fund. The House subsequently passed the Senate-amended version of the bill on September 8, 2017, and it was signed into law by President Trump the same day (P.L. 115-56), authorizing $15.3 billion in funding.[16]

Second Supplemental, October 2017 (P.L. 115-72, Division A)

On October 4, 2017, the Trump Administration requested an additional $12.7 billion for the Disaster Relief Fund (DRF). On October 12, the House passed H.R. 2266 with a further House amendment including $18.67 billion for the DRF (to the DHS Office of Inspector General for disaster audits), and also allowed some of that funding to be transferred to two other programs: $4.9 billion would go to FEMA's Disaster Assistance Direct Loan Program account, and $10 million to the Department of Homeland Security (DHS) Office of Inspector General for oversight of disaster-related activities. The bill subsequently was signed into law as P.L. 115-72 on October 26, 2017, authorizing $36.5 billion in funding. Of the $4.9 billion, up to $150 million is available for the cost of providing loans through the Advance of Non-Federal Share Program for the cost shares for Puerto Rico and the U.S. Virgin Islands related to Hurricanes Irma and Maria, and $1 million is for administrative expenses for the program.

---

[16] Ibid., p. 3.

Third Supplemental, February 2018 (P.L. 115-123, Division B)

The Trump Administration made a third supplemental appropriations request for disaster relief and recovery funding on November 17, 2017, seeking roughly $44.0 billion in additional funding. On February 7, 2018, the Senate took up H.R. 1892, with the Senate leadership adding S.Amdt. 1930. This became the Bipartisan Budget Act of 2018 (P.L. 115-123). Subdivision I of Division B of the amendment was titled "Further Additional Supplemental Appropriations for Disaster Relief Requirements Act, 2018" and included more than $84 billion in additional disaster assistance funding. The amended bill passed the House and was signed into law by President Trump as P.L. 115-123, authorizing $84.3 billion in funding.[17] Out of that amount, FEMA received an additional $23.5 billion in funding authority for the DRF.

P.L. 115-123 transferred up to $150 million of the $23.5 billion for DRF to the Disaster Assistance Direct Loan Program for costs related to Hurricanes Irma and Maria, of which $1 million may be used for administrative expenses for the program.

Funding for Power Restoration in Puerto Rico

In the Appendix, *Table A-1* shows FEMA's estimate of the allocations for the approximately $3.2 billion in funding for various electric power restoration activities in Puerto Rico under federal disaster supplemental appropriations.

According to FEMA, disaster charges related to Hurricanes Irma and Maria are being charged to P.L. 115-56 and P.L. 115-72. Thus, Disaster Declaration 4339 is being charged to P.L. 115-56 and P.L. 115-72. Once those funds are exhausted, funds from P.L. 115-123 will be used.[18] However, FEMA has also stated that there has been discussion about which supplemental will be eventually charged with the various activities, and it is possible that the allocation of funds could change in the future.

---

[17] Ibid., p.4.

[18] Email communication from Jerry Johnson, FEMA Congressional Affairs, June 19, 2018.

Preparation for the 2018 Hurricane Season

In recognition of the current fragility of the electric system in Puerto Rico, FEMA has left behind three large mobile power generation units, and charged the cost of the units to Disaster Declaration 4339.[19] FEMA has also left behind a number of peaking generation units to support power demands on the grid, and these are also charged to the same disaster declaration account.[20]

End of USACE Mission Assignment

The USACE's power restoration mission in Puerto Rico essentially ended on May 18, 2018, with approximately 99% of electric power customers having their power restored. Some USACE personnel remain to maintain the temporary generators still providing power on the island.

REBUILDING PUERTO RICO'S GRID

The grid emergency in Puerto Rico led to triage-type decisions, resulting in electric distribution poles and transmission towers that were marginally structurally sound being left in place in some instances. These poles and towers will almost certainly have to be replaced,[21] as the focus of the Commonwealth and the federal government turns from power

---

[19] See *Table A-1* in the Appendix, "MEPA 142." Consists of three "Mega" 30 Megawatt mobile diesel-fueled generators.

[20] See *Table A-1* in the Appendix, "MEPA 113." Consists of 20 diesel-fueled combustion turbine generators (of varying sizes) at eight locations in the PREPA system to support peak load operation during the hours of highest daily, weekly, or seasonal system electricity demands.

[21] "Electricity transmission towers installed specifically for temporary emergency restoration should be considered for replacement, potentially by monopoles; many of the round monopole structures survived the 2017 storms." U.S. Congress, Senate Committee on Energy and Natural Resources, *Full Committee hearing to Examine Puerto Rico's Electric Grid*, Written Testimony of Assistant Secretary Bruce J. Walker, 115th Cong., 2nd sess., May 8, 2018, p. 4. (hereinafter, Walker Testimony.)

restoration to rebuilding what has been described in the media as a "teetering" grid.[22]

While USACE's initial power restoration efforts were focused on repairing Puerto Rico's grid to "pre-storm" conditions, the poor state of the electrical system soon led to FEMA realizing that improvements would have to involve mobilizing the USACE to "rebuild the grid to U.S. code standards."[23]

When asked to what extent were improvements made to Puerto Rico's grid under authority granted by the Restoration of Damaged Facilities, 44 C.F.R. §206.226(d), FEMA responded as follows:

> To date, power restoration has been completed as "Emergency work." The materials used for emergency power restoration were designed to current codes and standards. Many areas received materials that will ultimately strengthen the system. For example, some towers that toppled over (that were rusted and not maintained) were replaced with new towers. While some infrastructure was replaced and even upgraded to code, Puerto Rico Electric Power Authority (PREPA) was not able to use the optimal materials or design to maximize resilience, because it was focused on emergency work. That will be done in the permanent work phase.[24]

Modernizing Puerto Rico's Grid

Approximately $2 billion in CDBG funds were made available to "enhance or improve" any electric power systems damaged by Hurricane Maria. To help guide the process of rebuilding Puerto Rico's grid, DOE reports that five national laboratories have collaboratively built a model of

---

[22] Michael Weissenstein, *Puerto Rico grid 'teetering' despite $3.8 billion repair job*, Associated Press, May 31, 2018, https://www.apnews.com/fa210cd1434c4d909e6030c7da884bce/Puerto-Rico-grid-'teetering'-despite-.

[23] U.S. Congress, Senate Committee on Homeland Security and Governmental Affairs, 2017 Hurricane Season: Oversight of the Federal Response, 115th Cong., 1st sess., October 31, 2017, pp. https://www.hsgac.senate.gov/hearings/2017-hurricane-season-oversight-of-the-federal-response.

[24] Email communication from Jerry Johnson, FEMA Congressional Affairs, July 10, 2018.

Puerto Rico's electricity system to test how to place microgrids, determine where to place power lines underground, and test siting of renewable energy projects where they can be sheltered from damage by extreme weather events.[25] DOE believes its modeling efforts can therefore help guide HUD and FEMA CDBG investments to improve the power grid in Puerto Rico.[26]

In June 2018, DOE issued a report focused on enhancing the resilience of Puerto Rico's electric grid with recommendations that it believes may help prioritize investments for transmission, distribution, new generation, energy storage, microgrids, and strategic power reserves.[27] This would allow potential impacts on other critical infrastructure such as the petroleum, natural gas, and telecommunications sectors to be estimated.

An enhanced and improved electricity system in Puerto Rico would likely include Smart Grid technologies to allow incorporation of more varied energy choices, both at consumer and electric utility levels.[28] DOE describes the Smart Grid as "an intelligent electricity grid—one that uses digital communications technology, information systems, and automation to detect and react to local changes in usage, improve system operating efficiency, and, in turn, reduce operating costs while maintaining high system reliability."[29]

However, the extent to which a Smart Grid would be fully deployed in Puerto Rico would likely depend on an evaluation of the potential benefits and costs of projected applications.[30] Cyber- and physical security would

---

[25] Walker Testimony, Near Term Modeling Support at p. 3.

[26] "DOE seeks to facilitate collaboration with PREPA as they plan future investments and determine where financial resources will be most beneficial in strengthening Puerto Rico's grid and increasing its resilience." Ibid.

[27] U.S. Department of Energy, *Energy Resilience Solutions for the Puerto Rico Grid*, June 2018, https://www.energy.gov/sites/prod/files/2018/06/f53/DOE%20Report_Energy%20Resilience%20Solutions%20for%20the%20PR%20Grid%20Final%20June%202018.pdf. (Hereinafter, DOERecs.)

[28] CRS Report R43742, *Customer Choice and the Power Industry of the Future*, by Richard J. Campbell.

[29] DOE, *Transforming the Nation's Electricity System: The Second Installment of the Quadrennial Energy Review*, January 2017, p. S-4, https://www.energy.gov/sites/prod/files/2017/02/f34/Quadrennial%20Energy%20Review--Second%20Installment%20%28Full%20Report%29.pdf.

[30] See CRS Report R45156, *The Smart Grid: Status and Outlook*, by Richard J. Campbell.

have to be a consideration in the design, construction, and operations of a modern grid incorporating Smart Grid technologies.[31]

Prospective Funding for Electric Power Systems

Section 21101 of the Bipartisan Budget Act of 2018 (P.L. 115-123) provided $28 billion in supplemental appropriations to HUD, under the Community Development Fund. Under the appropriation, $16 billion of the appropriation was to be available for declared disasters in 2017, with $11 billion for state and local governments affected by Hurricane Maria. Of this amount, up to $2 billion was made available (until expended) for "enhanced or improved electrical power systems."[32] Puerto Rico and the U.S. Virgin Islands are not specifically named as the intended recipients, and other states or local governments are also likely eligible for the grants.

Section 21210 of the act also places a requirement for FEMA (with assistance from DOE and the Governor of Puerto Rico, among others) to report on a 12- and 24-month economic and disaster recovery plan for electric power systems and grid restoration.

Reorganization of the Electric Power System

Puerto Rico's Governor Rosselló initially proposed privatizing and selling PREPA's assets in January 2018. This was followed in March 2018 with a plan submitted to Puerto Rico's Legislative Assembly with details of how the privatization of the utility would proceed.[33] One of the goals of

---

[31] CRS Report R43989, *Cybersecurity Issues for the Bulk Power System*, by Richard J. Campbell.

[32] The funds "shall be awarded directly to the State, unit of general local government" pursuant to the Robert T. Stafford Disaster Relief and Emergency Assistance Act (42 U.S.C. 5121 et seq.).

[33] "In January, Governor Rosselló announced a transformation of Puerto Rico's energy sector that is expected to take at least 12–18 months. The transformation envisions possibly engaging a private operator for the Island's transmission and distribution system through an effective mechanism (e.g., long term concession agreement) which is compliant with federal funding requirements and complimentary to the anticipated and planned reconstruction and

the plan was to use public-private partnerships to stabilize the electric system and lessen prices for electricity customers.[34] Details of how the privatization effort could accomplish these goals are unclear at this time.

There have been two previous efforts at privatization of a public utility in Puerto Rico. Efforts were made in the 1990s and early 2000s to privatize the Puerto Rico Aqueduct and Sewer Authority (PRASA), but "service quality deteriorated and prices for consumers increased, as did the agency's operational deficit."[35]

IMPROVING RELIABILITY IN PUERTO RICO

Power recovery efforts in Puerto Rico have focused primarily on restoring power to electricity customers. However, according to reports in the media, the speed of emergency restoration has taken precedent over "basic quality standards" for the work.[36] Puerto Rico is taking steps to adopt guidelines for electric systems maintenance used on the mainland.[37] DOE has recommended that Puerto Rico follow guidelines for electric

---

hardening of that system following last year's storms. This transformation also contemplates a private ownership or operation of existing PREPA assets as well as development of greenfield generation projects." U.S. Congress, Senate Committee on Energy and Natural Resources, *Full Committee Hearing to Examine Puerto Rico's Electric Grid*, Written Statement of Walter M. Higgins, Chief Executive Officer of the Puerto Rico Electric Power Authority, 115th Cong., 2nd sess., May 8, 2018, p. 3. (Hereinafter, Higgins Testimony.)

[34] Danica Coto, "Puerto Rico Gov Signs Bill to Privatize Power Company Assets," Associated Press, June 20, 2016, http://www.newsobserver.com/news/business/article21352 0879.html.

[35] Lara Merling, *Privatization Won't Fix Puerto Rico's Broken Power Utility*, North American Congress on Latin America, February 1, 2018, https://nacla.org/news/2018/02/01/ privatization-won%E2%80%99t-fix-puertorico%E2%80%99s-broken-power-utility.

[36] Michael Weissenstein, *Puerto Rico Power Grid Fragile Despite $3.8 billion in Repairs* May 31, 2018, https://www.elp.com/articles/2018/05/puerto-rico-power-grid-fragile-despite-3-8-billion-in-repairs.html?cmpid= enl_powergridinternational_energy_news_update_2018-06-01&pwhid=f06a907a9051105157975581c97f23cc38623bb57ced5874273487344d1048513f 597a70868ccd3831813e7fe2d1f10fe146dcc0f78e37d799dcf22bd9576140&eid=288971764& bid=2125048.

[37] "PREPA, with the support and guidance of FEMA and the United States Department of Energy, is working to harden the power grid and deploy the proper equipment and processes to meet power grid compliance standards. This initiative will continue to be deployed over the coming years and is expected to improve the reliability and efficiency of the system." Higgins Testimony, p. 3.

industry standards suggested by the Rural Utilities Service of the U.S. Department of Agriculture.[38] However, the extent to which the restored infrastructure in Puerto Rico is rebuilt to current electrical standards will be important to the modernization efforts that would follow, and much of this work may have to be redone.

Among the recommendations by DOE for improving electric reliability in Puerto Rico was the recommendation for an effective mutual assistance agreement to be in place.[39] Mutual aid agreements help utilities recover from power outage situations often caused by severe weather events, and are typically negotiated in anticipation of a future need. Under mutual assistance agreements for electric utilities, the utility receiving aid generally pays for the costs incurred by the utility providing aid in accordance with the assistance agreement.[40]

Congress gave the Federal Energy Regulatory Commission (FERC) authority to oversee the reliability of the bulk-power system under the Energy Policy Act of 2005 (P.L. 109-58). Reliability standards were added as Section 215 of the Federal Power Act[41] to help ensure the reliable operation of the bulk power system so that "instability, uncontrolled separation, or cascading failures" will not occur as a result of a sudden disturbance. FERC-approved reliability standards address programs ranging from vegetation clearances in electric transmission line rights-of-way, to policies and procedures for critical infrastructure protection of power plants and supporting facilities. However, FERC-approved reliability standards are only applicable to the continental United States,

[38] "USDA Rural Utilities Service (RUS) standards should be adopted where feasible and appropriate to standardize equipment and design, which will aid with replacement in both regular and emergency situations. USDA RUS standards govern the engineering and component specification of all voltage ranges of electrical transmission and distribution networks used by every rural electric utility in North America that borrows from RUS." DOERecs, p. 18.

[39] "The Governor and PREPA should immediately ensure that updated, effective mutual aid agreements are primed to quickly provide support during the next event." Walker Testimony, p. 4.

[40] David Ferris and Peter Behr, "Cash Crunch Slowed Puerto Rico's Appeal for Grid Help, CEO says," *E&E News*, October 9, 2017, https://www.eenews.net/stories/1060063083/print.

[41] 16 USC 12, Sections 824 to 824w.

and not to U.S. territories such as Puerto Rico.[42] To date, no legislation has been proposed to change this.

Building Electric System Resilience

With the power restoration effort almost finished, the next focus for authorities will likely be on rebuilding Puerto Rico's electricity transmission and distribution systems, as a modern system built to U.S. industry standards is more likely to survive extreme weather events. According to DOE, vegetation trimming and clearance requirements for utilities on the U.S. mainland for reliability purposes should be enforced on Puerto Rico, and are likely to reduce power outages caused by weather-related events.[43]

While electric system reliability and system resiliency are related, they differ both in scope and regulatory requirement. Reliability, according to the U.S. Department of Energy (DOE), is the ability of the system or its components to withstand instability, uncontrolled events, cascading failures, or unanticipated loss of system components.[44] Resilience, as defined by DOE, is the ability of a system or its components to adapt to changing conditions and withstand and rapidly recover from disruptions.[45]

Authorities in Puerto Rico were encouraged by DOE to define what system resilience entails for the island's power system.[46] This may combine elements of power generation modernization, incorporating

[42] 150 FERC ¶ 61,095.
[43] DOERecs, p. 20.
[44] U.S. Department of Energy, *Transforming the Nation's Electricity Sector: The Second Installment of the QER*, January 2017, https://www.energy.gov/sites/prod/files/2017/01/f34/Chapter%20IV%20Ensuring%20Electricity%20System%20Reliability%2C%20Security%2C%20and%20Resilience.p df.
[45] Ibid.
[46] "The report also notes where additional analysis is needed to more precisely articulate resilience-related, investment-grade suggestions regarding the design and specification of the electricity system in Puerto Rico." DOERecs, p. 7.

microgrids with hardening of infrastructure for the island's grid and communications systems.[47]

Electric System Resilience Must Be Planned

Once the backbone of Puerto Rico's electric system infrastructure is in place, then the major effort of making the system more resilient can follow. Enhancing resilience would likely require improvements that go beyond even rebuilding the electric power system according to standards in effect for U.S. mainland power utilities. Resilience may require additional improvements to the system aimed at better withstanding the effects of extreme weather events.

According to the DOE, there are no commonly used metrics for measuring grid resilience. Electric system resilience is not mandated by federal law, but the ability of the system to adapt to changing conditions and recover rapidly from disruptions is a key attribute of electric system reliability. A 2016 report from several of DOE's national laboratories focused on the potential for a changing environment, and the need to maintain a resilient power system.[48] The report identified risks ranging from weather events that disrupt transmission or distribution, to high impact, low frequency (HILF) risks such as catastrophic hurricanes. The report also built upon previous developments identified by DOE in the energy sector including growing potential threats from climate change, energy security, transitions from coal to natural gas generation, increased deployment of distributed and renewable generation, and rising

---

[47] Executive Office of the President, *Economic Benefits of Increasing Electric Grid Resilience to Weather Outages*, August 2013, https://www.energy.gov/sites/prod/files/2013/08/f2/Grid%20Resiliency%20Report_FINAL.pdf.

[48] Benjamin L. Preston, Scott N. Backhaus, and Mary Ewers, et al., *Resilience of the U.S. Electricity System: A Multi-Hazard Perspective*, DOE National Laboratories, August 2016, https://www.energy.gov/sites/prod/files/2017/01/f34/Resilience%20of%20the% 20 U.S.%20Electricity%20System%20A%20Multi-Hazard%20Perspective.pdf. (Hereinafter, GR.)

investments to modernize the energy grid.[49] While the report acknowledged that electric power systems "are currently well-equipped to effectively manage a broad range of threats," it recognized that some risks remain challenging and that "resilience efforts should shift toward these more complex risk management challenges." Some of the key risks to resilience identified in the study include the following:

- HILF threats associated with natural hazards (particularly weather or space weather) of historic intensity or large-scale physical, cyber, or electromagnetic attacks.
- Combined or blended threats associated with simultaneous exposure of the electric grid to one or more natural threats in combination with a physical or cyberattack.
- Threats that affect vulnerable components of the electricity system or that exceed critical thresholds. For example, distribution networks are often a weak link in the electric grid, but disruptions and outages associated with distribution are often localized.

The report provided a number of recommendations "to guide future decision-making to enhance resilience of the U.S. electricity system."[50] These recommendations included the following:

- Build a greater understanding of HILF events and capability to incorporate HILF threats into risk assessment. Scenario-based planning to explore multiple contingencies can be used to stress test the system and identify gaps in resilience.
- Develop a robust and scalable system of resilience metrics for the electricity system.
- Increase capacity to assess and manage risks and their uncertainties which may change over time and geographic areas. Future changes

---

[49] DOE, *The Quadrennial Energy Review*, January 2017, https://www.energy.gov/policy/initiatives/quadrennialenergy-review-qer.

[50] GR, p. 43.

in not only the climate, but also population, technology, and societal preferences have important implications for resilience.

- Institute policies and practices that can help to streamline assessment and decisionmaking while enhancing coordination and communication can be just as important to resilience as the development of robust infrastructure and assets.

How electric power systems incorporate resiliency into reliability planning will depend on their evaluation of risk to the system, and the financial and other resources available to system planners. Given the potential consequences of long-term electric power failures in Puerto Rico, Congress may consider further how various electric power systems incorporate resilience into reliability planning.

OTHER ISSUES GOING FORWARD

Puerto Rico's reliance on fossil fuels for power generation raises cost, reliability, and potential health issues going forward.

Before the 2017 hurricane season, coal and diesel fuel represented approximately 64% of fuel used for power generation. And the mega- and peaking electric combustion turbine generators left behind on Puerto Rico by FEMA are currently powered by burning diesel fuel. Coal and diesel fuel are expensive to import to Puerto Rico, and the reliance on diesel fuel for almost 50% of electricity on the islands has resulted in high prices, about 24 cents per Kilowatt-hour (kWh) for residential customers for power when compared to an average price of 13 cents per KWh on the U.S. mainland.[51] On the financial side, PREPA's inability to pay for fuel for power generation was a major complicating factor following Hurricane Maria.

[51] U.S. Energy Information Administration, *Why Puerto Rico's Electric Grid Stood No Chance Against Maria*, June 21, 2018, https://www.eia.gov/state/print.php?sid=RQ.

On the health side, among other factors (such as an increase in mold spores), emissions from diesel- and gasoline-fueled power generators of all sizes have been reported in the media as being linked to rising asthma rates in Puerto Rico since the 2017 storm season.[52]

Puerto Rico adopted a Renewable Portfolio Standard (RPS) in July 2010 (Act 82 of 2010) which mandated that PREPA supply increasing amounts of retail electricity sales from eligible "green energy" resources, peaking at 20% of retail sales by 2035. In 2014, concerns about PREPA's ability to follow through with this goal led to new legislation (Act 57 of 2014) that established a Puerto Rico Energy Commission and an office of consumer advocacy.[53] The Puerto Rico Energy Commission must have an independent regulatory role if RPS goals are to be met.[54]

Increasing renewable electric generation and requiring more energy storage may potentially reduce electricity costs, since most renewable electric technologies do not directly require a fossil fuel to produce power. However, the variability and intermittency of renewable generation means that energy storage, and traditional power generation capable of flexible, efficient operation (to increase or decrease power generation to support renewables) to provide reliable electricity would likely be needed. Increasing natural gas generation has been discussed by some as an option for Puerto Rico, since it is generally considered a cleaner-burning option to diesel fuel. But given Jones Act restrictions on importing liquefied natural gas to Puerto Rico, whether and how authorities may follow through with this option is unclear.[55]

---

[52] Ayana Bird, *Asthma Rates Rise in Puerto Rico Following Hurricane Maria*, June 18, 2018, https://www.colorlines.com/articles/asthma-rates-rise-puerto-rico-following-hurricane-maria.

[53] CRS Report R45023, *Repair or Rebuild: Options for Electric Power in Puerto Rico*, by Richard J. Campbell, Corrie E. Clark, and D. Andrew Austin.

[54] Doreen Hemlock, *Can Puerto Rico Really Depend on Microgrids and Renewables for Power?*, Renewable Energy World, December 15, 2018, https://www.renewableenergyworld.com/articles/2017/12/can-puerto-rico-really-dependon-microgrids-and-renewables-for-power.html.

[55] CRS Report R45006, *U.S. Liquefied Natural Gas (LNG) Exports: Prospects for the Caribbean*, by Michael Ratner et al.

RECENT LEGISLATION

In the 115th Congress, S. 1894 was introduced in September 2017, as a bill to exempt Puerto Rico from the coastwise laws of the United States. The bill would revise the coastwise laws, commonly known as the Jones Act, that govern domestic transportation of merchandise or passengers by vessels. The Jones Act requires that vessels transporting merchandise or passengers between Puerto Rico and other U.S. ports be built in the United States, at least 75% owned by U.S. citizens, and mostly crewed by U.S. citizens. Jones Act requirements are currently waived with respect to vessels transporting passengers between Puerto Rico and U.S. ports. This bill would permanently exempt vessels transporting merchandise between Puerto Rico and other U.S. ports from those requirements.

In the 115th Congress, the "Puerto Rico and Virgin Islands Equitable Rebuild Act of 2017" (S. 2165), introduced in November 2017, would provide for additional disaster-recovery assistance and other assistance to Puerto Rico and the U.S. Virgin Islands with respect to infrastructure, health care, agriculture, education, economic development, and environmental remediation, among other sectors. Specifically, regarding energy use in both territories, the bill would provide for the use of certain emergency assistance to rebuild electric grids, and would establish grant programs to promote energy efficiency and renewable energy.

In the 115th Congress, the "Puerto Rico and Virgin Islands Equitable Rebuild Act of 2018" (H.R. 4782), introduced in January 2018, would provide additional disaster-recovery assistance and other assistance to Puerto Rico and the U.S. Virgin Islands with respect to infrastructure, health care, agriculture, education, economic development, and environmental remediation, among other sectors. Specifically, regarding energy use in both territories, the bill would provide for the use of certain emergency assistance to rebuild electric grids, and would establish grant programs to promote energy efficiency and renewable energy.

APPENDIX. FEMA—PUERTO RICO TEMPORARY POWER RESTORATION OBLIGATIONS

Table A-1. FEMA Estimated Funding for Electric Power Restoration in Puerto Rico As of June 18, 2018

Date	Disaster Declaration Number	Applicant Name	Application Title	Amount Obligated[d] $	Mission Assignment $	Project Worksheet $
9/29/17	4339[a]	US Army Corps of Engineers (USACE) SAD16	Activation-Contracts/labor to execute mission	87,500,000	87,500,000	
10/08/17	4339	USACE SAD16	Contracts/labor to execute mission	489,500,000	489,500,000	
10/10/17	3384[b]	PREPA[c]	Emergency Protective Measures (EMP)	7,200,255		7,200,255
10/11/17	4339	PREPA	AEE001B EMP	45,000,000		45,000,000
10/14/17	4339	PREPA	PREPA—Emergency Power Plant Fuel[e]	42,800,000		42,800,000
10/15/17	4339	PREPA	PREPA—Emergency Power Plant Fuel[e]	128,400,000		128,400,000
10/28/17	4339	USACE	Contracts/labor to execute mission	450,000,000	450,000,000	
11/30/17	4339	USACE SAD16	Contracts/labor to execute mission	802,600,000	802,600,000	
12/30/17	4339 4339	PREPA PREPA	251E2—AEE002 EMP COBRA CONTRACT PREPA Emergency Power Plant Fuel de-obligation[e]	200,000,000 (171,200,000)		200,000,000 (171,200,000)
1/5/18	4339	USACE SAD16	Contracts/labor to execute mission	80,000,000	80,000,000	

Date	Disaster Declaration Number	Applicant Name	Application Title	Amount Obligated[d] $	Mission Assignment $	Project Worksheet $
3/6/18	4339	PREPA	314E1- EMP	200,000		200,000
3/9/18	4339	PREPA	251E2 – AEE002 EMP COBRA CONTRACT	745,429,800		745,429,800
3/16/18	4339	USACE SAD16	Contracts/labor to execute mission	44,000,000	44,000,000	
3/23/18	4339	USACE SAD16	Contracts/labor to execute mission	200,000,000	200,000,000	
5/1/18	3384	PREPA	EEPA004—EMP	10,625		10,625
5/17/18	4339	USACE	Net de-obligation (-$250,000,000 + $115,000,000)	(135,000,000)	(135,000,000)	
5/17/18	4339	USACE SAD33	Contracts/labor to execute mission	45,000,000	45,000,000	
5/20/18	4336[b]	PREPA	IEPA001—EMP	95,507		95,507
5/25/18	4339	PREPA	MEPA037—EMP	62,373		62,373
5/31/18	4339 4339	PREPA PREPA	MEPA113—EPM-Peaking Generator Units MEPA142—Mega Generator Purchases	130,897,179 58,500,000		130,897,179 58,500,000
6/1/18	3384	PREPA	EEPA005 EMP Distribution	49,962		49,962
6/2/18	4339	PREPA	MEPA063—EMP Canovanas	65,923		65,923
6/11/18	4339	USACE SAD16	De-obligate contract funds Total	(60,000,000) 3,191,111,623	(60,000,000) 2,003,600,000	1,187,511,623

Source: Email communication from Jerry Johnson, FEMA Congressional Affairs, June 18, 2018.

Notes: Columns may not add due to rounding.

[a] Declaration number 4339 is for Hurricane Maria.

[b] Emergency declaration number 3384 and Disaster declaration number 4336 are for Hurricane Irma.

[c] PREPA is the Puerto Rico Electric Power Authority (Autoridad de Energía Eléctrica, in Spanish).

[d] Funds made available to the State via electronic transfer following FEMA's final review and approval of Public Assistance projects (see https://www.fema.gov/ disaster-page-definitions).

[e] The emergency fuel funds were returned to FEMA, and the funds de-obligated.

Funds listed under "Project Worksheet" in *Table A-1* are amounts associated with FEMA's Public Assistance program.[56] Funds listed under Mission Assignment refer to the work orders issued by FEMA to another federal agency directing the completion of a specific task.[57]

[56] As a public utility, PREPA is an eligible applicant that can receive federal assistance through the Federal Emergency Management Agency (FEMA). In particular, FEMA provides grant assistance through the Public Assistance Grant Program (PA Program) for the repair, restoration, and replacement of public facilities, as defined by law, in states and communities that have received a major or emergency disaster declaration through the Robert T. Stafford Disaster Relief and Emergency Assistance Act (Stafford Act, P.L. 93-288, as amended). For an overview of the PA Program, see CRS Report R43990, *FEMA's Public Assistance Grant Program: Background and Considerations for Congress*, by Jared T. Brown and Daniel J. Richardson, *FEMA's Public Assistance Grant Program: Background and Considerations for Congress*.

[57] FEMA, *Federal Agencies Providing Disaster Assistance*, 2018, https://www.fema.gov/federal-agencies-providingdisaster-assistance.

In: Hurricanes, Wildfires and Flooding
Editor: Lydie Yohan
ISBN: 978-1-53616-362-9

Chapter 6

U.S. VIRGIN ISLANDS RECOVERY: STATUS OF FEMA PUBLIC ASSISTANCE FUNDING AND IMPLEMENTATION*

United States Government Accountability Office

WHY GAO DID THIS STUDY

In September 2017, two major hurricanes—Irma and Maria—struck the USVI, causing billions of dollars in damage to its infrastructure, housing, and economy. FEMA—a component of the Department of Homeland Security—is the lead federal agency responsible for assisting the USVI as it recovers from these natural disasters. Among other responsibilities, FEMA administers the Public Assistance program in partnership with the USVI territorial government, providing the USVI grant funding for response and recovery activities, including debris

* This is an edited, reformatted and augmented version of United States Government Accountability Office; Report to Congressional Requesters, Publication No. GAO-19-253, dated February 2019.

removal efforts, life-saving emergency protective measures, and the repair, replacement, or restoration of public infrastructure.

GAO was asked to review the federal government's response and recovery efforts related to the 2017 hurricanes. This chapter describes (1) the status of FEMA's Public Assistance program funding provided to the USVI in response to the 2017 hurricanes as of October 1, 2018, and (2) the USVI's transition to implementing the Public Assistance alternative procedures in the territory. GAO reviewed program documents and data on obligations and expenditures as of October 1, 2018, and interviewed officials from FEMA and the USVI regarding the Public Assistance program specifically and disaster recovery efforts more generally. GAO also conducted site visits to the USVI islands of St. Croix, St. Thomas, and St. John.

GAO is not making any recommendations in this chapter, but will continue to monitor the progress of the USVI's recovery as part of its ongoing work.

What GAO Found

The Federal Emergency Management Agency (FEMA) obligated more than $1.4 billion in grant funding for Public Assistance projects in the U.S. Virgin Islands (USVI) as of October 1, 2018, in response to the 2017 hurricanes. FEMA obligated about $873.8 million for emergency work—debris removal activities and emergency measures to lessen the immediate threat to life, public health, and safety—and about $516.3 million for permanent work—including the repair or replacement of public infrastructure such as roads, electrical utilities, and schools. For example, FEMA obligated about $101 million for the purchase and installation of modular units to be used as temporary classrooms and other facilities while permanent school buildings are repaired or replaced. FEMA's obligations for permanent work also included funding for hazard mitigation measures to reduce the risk of damage during future storms—for example, by replacing wooden utility poles with composite fiberglass poles (see Figure).

FEMA and the USVI are transitioning from using the standard Public Assistance program in the territory to using the Public Assistance alternative procedures program. Unlike in the standard Public Assistance program where FEMA will fund the actual cost of a project, the alternative procedures allow awards to be made on the basis of fixed-cost estimates to provide financial incentives for the timely and cost-effective completion of permanent work projects. FEMA and USVI officials stated that the alternative procedures will give the USVI more flexibility in determining when and how to fund projects and provide an opportunity to repair and rebuild the USVI's critical services infrastructure—such as its education system and electrical grid—so it meets industry standards without regard to pre-disaster condition. As of November 2018, FEMA and USVI officials were discussing the process for developing projects under the Public Assistance alternative procedures. GAO will continue to monitor the USVI's plans for using the alternative procedures as part of its broader review assessing the USVI's disaster recovery efforts and will issue a follow-on report later this year.

February 25, 2019
Congressional Requesters

In the span of 14 days in September 2017, two Category 5 hurricanes— Irma and Maria—struck the territory of the U.S. Virgin Islands (USVI), causing billions of dollars in damage to its infrastructure, housing, and economy.[1] These disasters exacerbated an already difficult

[1] The National Oceanic and Atmospheric Administration (NOAA) measures hurricanes on a scale from 1 to 5 with a Category 1 being the least intense and a Category 5 being the most intense. NOAA defines a Category 5 hurricane as one with winds above 157 miles per hour. Both Hurricane Irma and Hurricane Maria made landfall on the U.S. Virgin Islands (USVI) as Category 5 hurricanes.

financial situation in the USVI, which was operating under severe fiscal constraints prior to the storms.[2]

The Federal Emergency Management Agency (FEMA)—a component of the Department of Homeland Security (DHS)—is the lead federal agency responsible for disaster preparedness, response, and recovery, which includes assisting the USVI as it recovers from these natural disasters.[3] Among other responsibilities, FEMA administers the Public Assistance program in the USVI through a partnership with the territorial government and provides grant funding for a wide range of eligible response and recovery activities. These activities include debris removal efforts, life- saving emergency protective measures, and the repair, replacement, or restoration of disaster-damaged publicly-owned facilities, electrical utilities, roads and bridges, and more. Further, in July 2018, FEMA approved the USVI government's request to use the Public Assistance alternative procedures program in the territory that offers financial incentives for the timely and cost-effective completion of permanent work.

You asked us to review the federal government's response and recovery efforts related to the 2017 hurricanes. As part of that effort, we have several reviews ongoing of FEMA's and the USVI government's disaster recovery planning, oversight, and execution efforts in the USVI. In this chapter, we discuss:

1) the status of FEMA Public Assistance program funding provided to the USVI in response to the 2017 hurricanes as of October 1, 2018— more than 1 year after hurricanes Irma and Maria impacted the territory; and
2) the USVI's transition to implementing FEMA's Public Assistance alternative procedures program in the territory.

---

[2] The USVI's public debt nearly doubled between fiscal years 2005 and 2015, reaching $2.6 billion and a debt to Gross Domestic Product ratio of 72 percent. GAO, *U.S. Territories: Public Debt Outlook*, GAO-18-160 (Washington, D.C.: Oct. 2, 2017).

[3] See 6 U.S.C. § 313.

Another review will examine disaster recovery efforts in the USVI and we plan to issue that report later this year. In addition, we are conducting a broader body of work covering various disaster response and recovery issues, including disaster recovery in impacted states in the contiguous United States and Puerto Rico, and will be reporting on these issues over the next year.

For this chapter, we reviewed relevant documentation on FEMA's Public Assistance program—including the April 2018 *Public Assistance Program and Policy Guide* and the September 2018 *Public Assistance Alternative Procedures Permanent Work Guide* for the USVI.[4] We also obtained and analyzed data from FEMA's Emergency Management Mission Integrated Environment and Integrated Financial Management Information System on Public Assistance program obligations and the USVI's expenditures as of October 1, 2018.[5]

An obligation is a definite commitment that creates a legal liability of the government for the payment of goods and services ordered or received. For the purposes of this chapter, obligations represent the amount of grant funding FEMA provided through the Public Assistance program for specific projects in the USVI. An expenditure is an amount paid by federal agencies, by cash or cash equivalent, during the fiscal year to liquidate government obligations. For the purposes of this chapter, an expenditure represents the actual spending by the USVI government of money obligated by the federal government. We interviewed FEMA officials, including those stationed in the USVI, as well as officials from the USVI's Governor's office, the USVI's Territorial Emergency Management Agency, the USVI's Department of Finance, and contractor personnel regarding the implementation of the Public Assistance program in the territory. Further, in May and October of 2018, we conducted site visits to the USVI islands of St. Croix, St. Thomas, and St. John to observe

---

[4] *Public Assistance Alternative Procedures (Section 428) Permanent Work Guide FEMA- 4340-DR-VI* (September 15, 2018).

[5] These data do not include obligations and expenditures for, among other federal disaster assistance programs, direct federal mission assignments, in which a federal agency is tasked with providing eligible emergency work or debris removal services to a territory or state, or for other categories of mission assignments.

hurricane-damaged areas and eligible Public Assistance projects, as well as disaster assistance funding from other federal departments.

We assessed the reliability of the Emergency Management Mission Integrated Environment and the Integrated Financial Management Information System data by reviewing existing information about these systems, interviewing data users and managers responsible for these data, and cross-checking data across sources to ensure consistency. Based on these steps we determined these data to be sufficiently reliable for the purposes of this chapter.

We conducted this performance audit from October 2018 through February 2019 in accordance with generally accepted government auditing standards. Those standards require that we plan and perform the audit to obtain sufficient, appropriate evidence to provide a reasonable basis for our findings and conclusions based on our audit objectives. We believe that the evidence obtained provides a reasonable basis for our findings and conclusions based on our audit objectives.

BACKGROUND

Hurricanes Irma and Maria

In September 2017, two Category 5 hurricanes struck the USVI, causing catastrophic damage across the entire territory and neighboring Caribbean islands.[6] On September 6, 2017, Hurricane Irma struck St.

Thomas and St. John and on September 19, 2017, Hurricane Maria struck St. Croix (see Figure 1).[7]

---

[6] The USVI is the easternmost territory of the United States and consists of four islands (St. Thomas, St. Croix, St. John, and Water Island). The USVI is located 1,115 miles southeast of Miami and 70 miles east of San Juan, Puerto Rico (see Figure 1).

[7] We will report on the status of FEMA Public Assistance funding and recovery planning efforts in Puerto Rico as a result of hurricanes Irma and Maria in a separate report later this year.

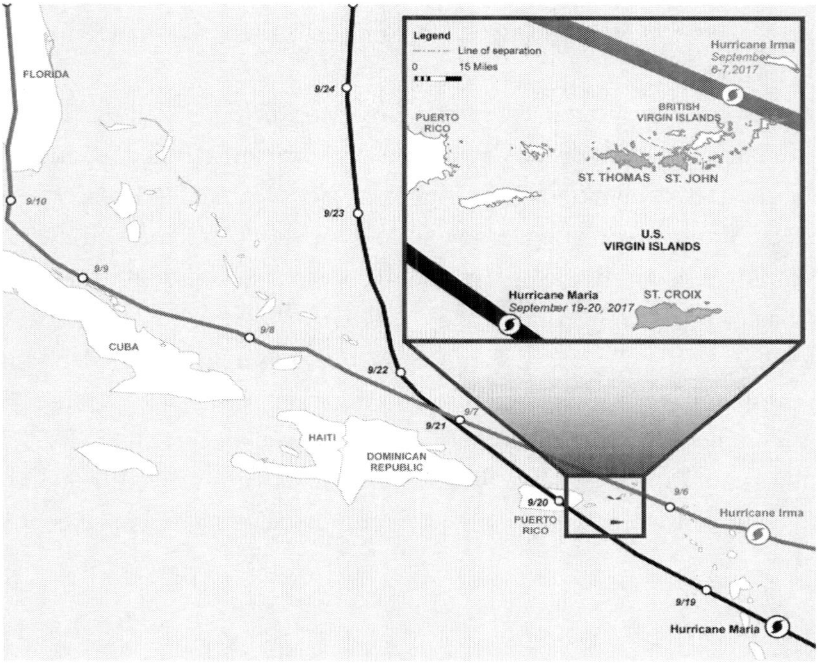

Source: GAO analysis of National Oceanic and Atmospheric Administration data; Map Resources (map). | GAO-19-253.

Figure 1. Path of Hurricanes Irma and Maria through the U.S. Virgin Islands.

The storms severely damaged the territory's critical infrastructure, devastating more than 90 percent of aboveground power lines and shutting down electricity and telecommunications for months.[8] Further, 52 percent of the territory's housing units were damaged, ports and airports were closed for weeks, and hundreds of thousands of tons of debris were generated, often blocking roads and making transportation hazardous. In addition, according to a September 2018 report from the USVI Hurricane Recovery and Resilience Task Force, the territory's economic activity—especially tourism—was severely reduced in the months following the

[8] We will report our observations on federal support for electricity grid restoration in Puerto Rico and the USVI as a result of the 2017 hurricanes in a separate report later this year.

storms, leading to job losses and a total estimated economic impact of $1.54 billion.[9]

In response to the request of the Governor of the USVI, the President declared a major disaster the day after each hurricane struck the territory.[10] Major disaster declarations can trigger a variety of federal response and recovery programs for government and nongovernmental entities and households and individuals, including assistance through the Public Assistance program.[11] Under the *National Response Framework* and *National Disaster Recovery Framework*, DHS is the federal department with primary responsibility for coordinating disaster response and recovery, and within DHS, FEMA has lead responsibility. The Administrator of FEMA serves as the principal adviser to the President and the Secretary of Homeland Security regarding emergency management.[12]

---

[9] In October 2017, the Governor of the USVI called for the USVI Hurricane Recovery and Resilience Task Force to draft a report assessing the USVI's hurricane response and guiding its efforts during the rebuilding process. The report was released in September 2018 and details recovery initiatives across 14 sectors, including the economy, energy, communications, transportation, and more. The report can be found at https://www.usvihurricanetaskforce.org/.

[10] In accordance with the Robert T. Stafford Disaster Relief and Emergency Assistance Act (Stafford Act), as amended, the President of the United States may declare that a major disaster or emergency exists in response to a governor's or tribal chief executive's request if the disaster is of such severity and magnitude that effective response is beyond the capabilities of a state, tribe, or local government and federal assistance is necessary. See 42 U.S.C. § 5170. Prior to a major disaster or emergency declaration, a preliminary damage assessment is typically used to determine the impact and magnitude of damage and the resulting needs of individuals, businesses, the public sector, and the community as a whole. The territory then generally uses the information collected as the basis for the governor's request for a major disaster or emergency declaration. However, based on the severity and magnitude of hurricanes Irma and Maria, the FEMA Regional Administrator, upon consultation with the USVI, determined supplemental federal assistance was necessary before joint federal, state, and local government preliminary damage assessments were completed.

[11] *Presidential Policy Directive-8 National Preparedness* (PPD-8) establishes a national preparedness system made of an integrated set of guidance, programs, and processes designed to strengthen the security and resilience of the United States through systematic preparation for the natural and human-caused threats that pose the greatest risk. This system breaks preparedness activities into five different lines of effort—prevention, protection, mitigation, response, and recovery—each of which requires a separate planning framework.

[12] 6 U.S.C. § 313(c)(4).

FEMA's Public Assistance Program

FEMA's Public Assistance program provides funding to state, territorial, local, and tribal governments as well as certain types of private nonprofit organizations to assist with responding to and recovering from major disasters or emergencies.[13] As shown in Figure 2, Public Assistance program funds are categorized broadly as "emergency work" or "permanent work." Within these broad categories are separate subcategories. In addition to the emergency work and permanent work categories, the program includes category Z, which represents indirect costs, administrative expenses, and other expenses a recipient or subrecipient incurs in administering and managing the Public Assistance program that are not directly chargeable to a specific project.

FEMA's Public Assistance program also provides funding for cost-effective hazard mitigation measures to reduce or eliminate the long-term risk to people and property from future natural and man-made disasters and their effects.[14] Specifically, FEMA provides funding for hazard mitigation measures in conjunction with the repair of disaster-damaged facilities to enhance their resilience during future disasters. For example, a community that had a fire station damaged by a disaster could use Public Assistance funding to repair the facility and incorporate additional measures such as installing hurricane shutters over the windows to mitigate the potential for future damage.

---

[13] The Public Assistance program represents the largest share of the Disaster Relief Fund, which is the primary source of federal disaster assistance for state and local governments when a disaster is declared. The Disaster Relief Fund is appropriated no-year funding, which allows FEMA to fund, direct, coordinate, and manage response and recovery efforts—including certain efforts by other federal agencies and state and local governments, among others—associated with domestic disasters and emergencies.

[14] FEMA may fund hazard mitigation measures related to the damaged facilities receiving Public Assistance funding pursuant to section 406 of the Stafford Act, as amended. 42 U.S.C. § 5172; 44 C.F.R. § 206.226. FEMA Public Assistance policy allows three different methods to test for cost-effective hazard mitigation. First, under the 15 percent rule, hazard mitigation measures may amount to up to 15 percent of the total eligible cost of repair work on a project. Second, certain hazard mitigation measures that have been predetermined to be cost-effective may qualify under the 100 percent rule, which permits the hazard mitigation as long as it does not exceed 100 percent of the eligible cost of the repair work on a project. Third, for measures that exceed eligible costs, the territory must demonstrate that the measure is cost-effective through an acceptable cost/benefit analysis methodology.

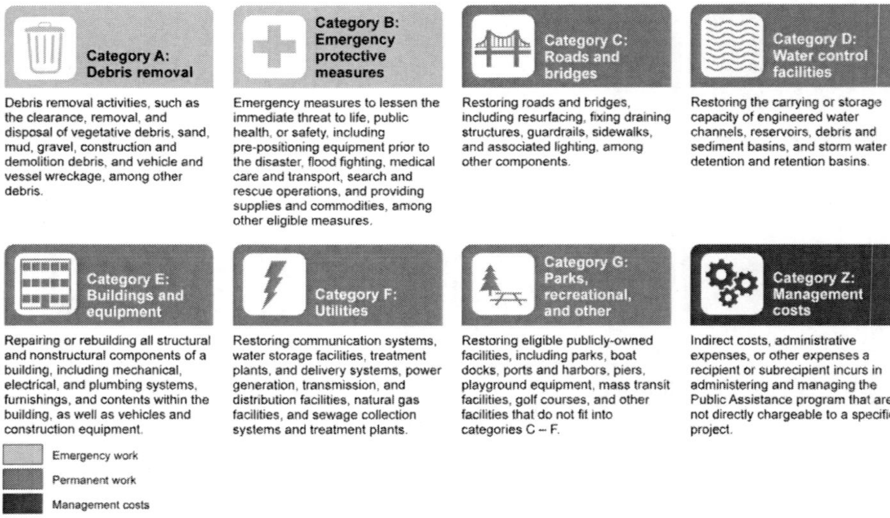

Source: Federal Emergency Management Agency. | GAO-19-253.

Figure 2. Federal Emergency Management Agency's Public Assistance Program Categories of Work.

Once the President has declared a disaster, FEMA, the state or territorial government (the recipient), and local or territorial entities (the subrecipient) work together to develop damage assessments and formulate project worksheets for eligible projects. Project worksheets detail the scope of work and estimated cost for repairing or replacing disaster-damaged infrastructure as well as any hazard mitigation measures that may help to increase the resilience of this infrastructure during future disasters. After a project has completed FEMA's review process and is approved, FEMA obligates funding for the project by placing money into an account where the recipient has the authority to draw down—or withdraw—funding to pay the subrecipient for eligible work upon completion (see Figure 3).

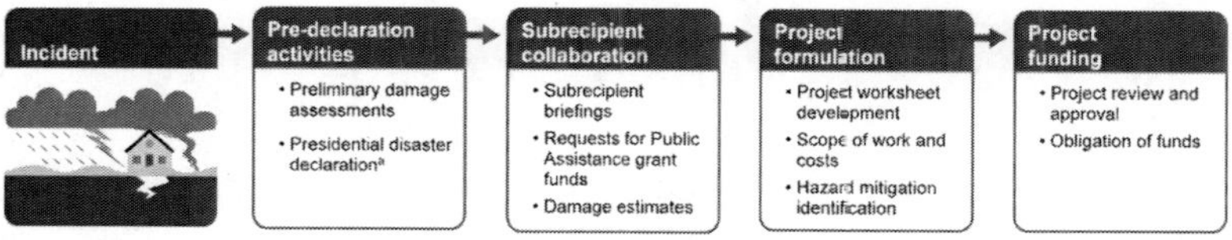

Source: GAO review of FEMA documents; Art Explosion (clip art). | GAO-19-253.

[a]In accordance with the Robert T. Stafford Disaster Relief and Emergency Assistance Act (Stafford Act), as amended, the President of the United States may declare that a major disaster or emergency exists in response to a governor's or tribal chief executive's request if the disaster is of such severity and magnitude that effective response is beyond the capabilities of a state, tribe, or local government and federal assistance is necessary. See 42 U.S.C. § 5170.

Figure 3. Federal Emergency Management Agency's (FEMA) Public Assistance Program Application Process.

In addition, a state or territorial governor may designate a governor's authorized representative (GAR) to oversee all aspects of disaster assistance, including Public Assistance funding.[15] Specifically, the GAR is responsible for ensuring compliance with program requirements by providing oversight into how goods and services are procured for projects, such as construction materials or modular school units. The GAR also confirms that subrecipients submit complete documentation demonstrating that all work completed is in accordance with a project's approved scope of work and Public Assistance program requirements.

The GAR then approves the paperwork and the recipient can draw down funding from the account holding the obligations to reimburse subrecipients for completed work. When a project has been completed, FEMA conducts a close-out process to certify that all work has been completed and reconciles the actual cost incurred. If the actual cost of the completed work is greater than the amount of money FEMA obligated for the project, FEMA will reimburse the subrecipient for these additional costs.

---

[15] 44 C.F.R. §§ 206.2(a)(13) and 206.41(d). The GAR is the designated individual responsible for administering federal disaster assistance programs on behalf of the state, local, tribal, or territorial government and ensuring compliance with all applicable laws, regulations, and FEMA guidance.

FEMA's Public Assistance Alternative Procedures Program

The Sandy Recovery Improvement Act of 2013 authorized the use of alternative procedures in administering the Public Assistance program, thereby providing new flexibilities to FEMA, states, territories, and local governments for debris removal, infrastructure repair, and rebuilding projects using funds from this program.[16] Unlike in the standard Public Assistance program where FEMA will fund the actual cost of a project, the Public Assistance alternative procedures allow awards for permanent work projects to be made on the basis of fixed-cost estimates to provide financial incentives for the timely and cost-effective completion of work.

Under these procedures, if the actual cost of the project exceeds the fixed-cost estimate agreed upon by FEMA and the recipient, the recipient or subrecipient is responsible for the additional costs at the time of the close-out process. However, if the actual cost of completing eligible work for a project is below the estimate, the recipient may use the remaining funds for additional cost-effective hazard mitigation measures to increase the resilience of public infrastructure. In addition, these funds may also be used for activities that improve the recipient's or subrecipient's future Public Assistance operations or planning.

[16] The Sandy Recovery Improvement Act of 2013 amended the Stafford Act by adding Section 428, which authorized FEMA to approve Public Assistance program projects under the alternative procedures provided by that section for any presidentially-declared major disaster or emergency. This section further authorized FEMA to carry out the alternative procedures as a pilot program until FEMA promulgates regulations to implement this section. Pub. L. No. 113-2, div. B, § 1102(2), 127 Stat. 39, amending Pub. L. No. 93-288, tit. IV, § 428 (codified at 42 U.S.C. § 5189f). The stated goals of the alternative procedures are to reduce the costs to the federal government, increase flexibility in the administration of the Public Assistance program, expedite the provision of assistance under the program, and provide financial incentives for recipients of the program for the timely and cost-effective completion of projects.

FEMA Had Obligated More Than $1.4 Billion and the USVI Had Expended About $587 Million in Public Assistance Funding as of October 1, 2018

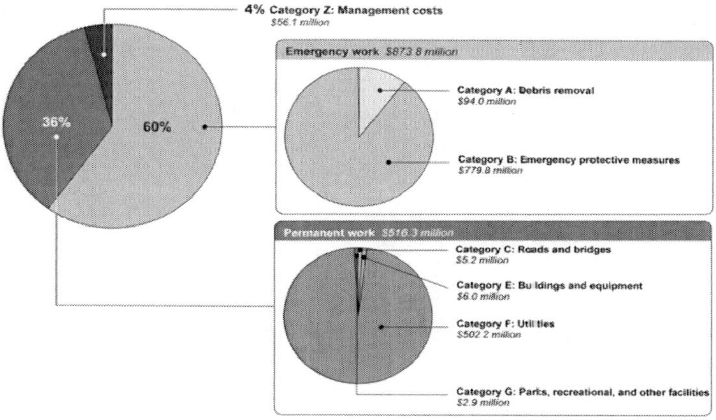

Source: Federal Emergency Management Agency's Emergency Management Mission Integrated Environment. | GAO-19-253.

Note: As of October 1, 2018, FEMA did not have any category D Public Assistance projects in the U.S. Virgin Islands. Emergency protective measures are activities to lessen the immediate threat to life, public health, or safety such as search and rescue operations and providing medical care and transport. Management costs are indirect costs, administrative expenses, and other expenses a recipient or subrecipient incurs in administering and managing the program that are not directly chargeable to a specific project.

[a]These data do not include obligations and expenditures for, among other federal disaster assistance programs, direct federal mission assignments, in which a federal agency is tasked with providing eligible emergency work or debris removal services to a territory or state, or for other categories of mission assignments.

Figure 4. Federal Emergency Management Agency's (FEMA) $1.4 Billion in Public Assistance Funding Obligated to the U.S. Virgin Islands by Category, October 1, 2018[a].

As of October 1, 2018, FEMA had obligated more than $1.4 billion through the standard Public Assistance program for 475 projects across the

USVI.[17] As shown in Figure 4, FEMA obligated funding for both emergency work and permanent work projects.

As of October 1, 2018, of the more than $1.4 billion FEMA obligated, the USVI had expended approximately $586.9 million—about 41 percent of total Public Assistance program obligations to the USVI—to reimburse subrecipients for completed work.[18] Of this $586.9 million, the USVI had expended about $532.8 million (91 percent) for emergency work projects in categories A and B and $49.1 million (8 percent) for permanent work projects in categories C through G.[19] The majority of FEMA's obligations and the funding the USVI expended as of October 1, 2018, are for emergency work because these projects began soon after the disasters struck and focused on debris removal and providing assistance to address immediate threats to life and property. In contrast, permanent work projects take time to identify, develop, and ultimately complete as they represent the longer-term repair and restoration of public infrastructure.[20]

---

[17] FEMA data presented in this chapter on obligations and expenditures in the USVI represent funding provided through FEMA's standard Public Assistance program. These data do not include obligations and expenditures for, among other federal disaster assistance programs, direct federal mission assignments, in which a federal agency is tasked with providing eligible emergency work or debris removal services to a territory or state, or for other categories of mission assignments. The GAR monitors and publicly reports on the status of Public Assistance program funding in the territory, among other disaster assistance programs, through the USVI Public Finance Authority website at http://www.usvipfa.com/hurricanehome.html.

[18] The Governor of the USVI designated two territorial entities to implement and oversee the Public Assistance program in the USVI—the USVI Territorial Emergency Management Agency as the programmatic manager and the Commissioner of the USVI's Department of Finance as the GAR. The USVI also hired two contractors to help identify and develop projects and provide oversight over Public Assistance funding draw downs. As the USVI does not have local (e.g., city or county) governance structures, decisions regarding the Public Assistance program and its implementation are made at the territorial government level and do not require input or approval from additional governmental entities. Further, during territorial emergencies like the 2017 hurricanes, the Governor of the USVI has the authority to make decisions regarding how disaster recovery activities will be implemented in the territory, according to USVI officials.

[19] The USVI also expended about $5.1 million for management costs under category Z. As noted above, this category represents indirect costs, administrative expenses, and other expenses a recipient or subrecipient incurs in administering and managing the program that are not directly chargeable to a specific project.

[20] The federal cost share—or the percentage of the total cost FEMA is responsible for obligating for a given project—may differ between emergency work and permanent work projects. In the USVI, the federal cost share for emergency work projects (categories A and B) was 100 percent until September 2018 and 90 percent thereafter, while the federal cost share for

Emergency Work

Of the more than $1.4 billion FEMA had obligated as of October 1, 2018, about $873.8 million (60 percent) was obligated for 322 emergency work projects in Public Assistance categories A and B.

- *Category A: Debris Removal.* FEMA obligated about $94.0 million for 71 projects focused on debris removal activities across the territory.[21]
  For example, FEMA obligated $45.0 million to the USVI Department of Public Works for territory-wide debris removal efforts and $39.1 million to the USVI Water and Power Authority for these activities in St. Croix (see Figure 5). Of the $94.0 million FEMA obligated for debris removal, the USVI had expended about $54.6 million (58 percent) as of October 1, 2018.
- *Category B: Emergency Protective Measures.* FEMA obligated about $780 million for 251 projects focused on emergency measures. For example, FEMA obligated about $187 million for the Sheltering and Temporary Essential Power program, which is intended to provide essential repairs or restore power to private residences to allow affected individuals to return or remain in their homes, thereby reducing the demand for other shelter options. In addition, FEMA obligated approximately $101 million for the purchase and installation of modular units to be used as temporary classrooms and other facilities while permanent school buildings are repaired or replaced (see Figure 6). Of the $780 million FEMA obligated for emergency protective measures, the USVI had expended about $478 million (61 percent) as of October 1, 2018.

---

permanent work projects (categories C through G) is 90 percent. The USVI is responsible for funding the percentage of emergency work and permanent work projects FEMA does not cover.

[21] Debris removal is the first step in the disaster recovery process, allowing communities to restore accessibility to public services and areas, while ensuring public health and safety in the aftermath of a disaster. Following a major disaster, FEMA provides funding assistance for the removal of debris and wreckage from publicly and privately-owned lands and waters when such removal is in the public interest. See 42 U.S.C. § 5173.

A drop-off and storage area in St. Croix, U.S. Virgin Islands for organic debris awaiting transportation to the continental United States for disposal.

Source: GAO. | GAO-19-253.

Figure 5. Organic Debris Storage Area in St. Croix, U.S. Virgin Islands.

Source: GAO. | GAO-19-253.

Figure 6. Two large modular units to provide indoor common areas, including a cafeteria. Construction of Temporary School Facilities in St. Croix, U.S. Virgin Islands.

Permanent Work

Of the more than $1.4 billion in Public Assistance funding FEMA had obligated as of October 1, 2018, about $516.3 million (36 percent) was obligated for 153 permanent work projects across categories C through G. These permanent work projects include about $349.4 million for cost-effective hazard mitigation measures aimed at reducing the future risk of disaster-damaged facilities in conjunction with their repair.[22] Further, of the $516.3 million FEMA obligated for permanent work in the USVI, approximately $500.4 million—or 97 percent of all permanent work obligations—was obligated to the USVI Water and Power Authority for the permanent repair of electrical distribution systems and other utilities across the territory.

- *Category C: Roads and Bridges.* FEMA obligated about $5.2 million for 35 projects focused on repairing roads and bridges in the territory, 18 of which included hazard mitigation measures totaling about $1.5 million. For example, FEMA obligated about $410,000 for one project to repair a road on St. Thomas damaged by floodwaters. This project included approximately $227,000 for hazard mitigation measures, such as replacing the damaged road surface with reinforced concrete and building a retaining wall. As of October 1, 2018, the USVI had not expended funding in this category.
- *Category D: Water Control Facilities.* As of October 1, 2018, FEMA did not have any projects in this category. According to FEMA officials, the USVI does not have water control infrastructure that would fall under category D, such as dams, levees, or berms.
- *Category E: Buildings and Equipment.* FEMA obligated $6.0 million for 77 projects focused on repairing and rebuilding damaged buildings and equipment, 16 of which included hazard mitigation measures totaling about $1.8 million. For example, FEMA obligated about $1.5 million to repair damage to the airport terminal building in St. Thomas—a project where hazard mitigation measures comprised 87 percent of the project's total cost (see Figure 7). These measures include replacing the

[22] FEMA has the authority to provide Public Assistance funding for cost-effective hazard mitigation measures to reduce or eliminate the long-term risk to people and property from natural hazards and their effects. See 42 U.S.C. § 5172; 44 C.F.R. § 206.226.

terminal's roof with materials designed to withstand higher wind speeds to increase the building's resilience during future storms. Of the $6.0 million FEMA obligated for category E, the USVI had expended about $148,000 (2.5 percent) as of October 1, 2018.

Source: GAO. | GAO-19-253.

Figure 7. Segments of the roof at the Cyril E. King International Airport in St. Thomas, U.S. Virgin Islands with blue plastic sheeting awaiting permanent repairs. Roof Damage at Cyril E. King International Airport in St. Thomas, U.S. Virgin Islands.

- *Category F: Utilities.* Of the $516.3 million FEMA obligated for permanent work projects, $502.2 million (97 percent) was obligated for 15 projects focused on repairing utilities, 7 of which included hazard mitigation measures totaling about $346.0 million.[23] For example, FEMA obligated $286.1 million and $50.2 million for permanent electrical distribution system repairs in St. Croix and St. John, respectively. This includes replacing damaged wooden utility poles with more resilient composite fiberglass poles that can withstand 200 mile per hour winds as well as power transmission lines and transformers (see Figure 8). Of the $502.2

[23] As of October 1, 2018, 99 percent of all funding for hazard mitigation measures in the USVI was for projects in Public Assistance category F.

million FEMA obligated for category F, the USVI had expended about $49.0 million (10 percent) as of October 1, 2018.

Source: GAO. | GAO-19-253.

Figure 8. Wooden utility poles for electricity distribution are being replaced by composite fiberglass poles that can withstand 200 mile per hour winds. Composite Fiberglass Utility Poles in St. John, U.S. Virgin Islands.

- *Category G: Parks, Recreational, and Other Facilities.* FEMA obligated about $2.9 million for 26 projects focused on repairing parks, playgrounds, and other recreational facilities, 1 of which included hazard mitigation measures. Specifically, FEMA obligated about $453,000 to repair the Lindbergh Park and Water Playground in St. Thomas—a project that included about $18,000 for hazard mitigation measures. As of October 1, 2018, the USVI had not expended funding in this category.

Future Projects

In addition to the more than $1.4 billion in Public Assistance funding FEMA had obligated as of October 1, 2018, FEMA expected to review an additional 900 future projects for eligibility representing an estimated

$779.4 million in potential funding.[24] Of this estimated total amount, FEMA anticipates $128.5 million (16 percent) in costs for future emergency work projects and $650.9 million (84 percent) in costs for future permanent work projects.[25]

FEMA AND THE USVI ARE TRANSITIONING FROM THE STANDARD PUBLIC ASSISTANCE PROGRAM TO THE PUBLIC ASSISTANCE ALTERNATIVE PROCEDURES PROGRAM

In July 2018, FEMA approved a June 2018 request from the Governor of the USVI to transition to using the Public Assistance alternative procedures program for permanent work in the territory.[26] The alternative procedures provide new flexibilities to FEMA and the USVI that are not available through the standard Public Assistance program. In September 2018, FEMA issued the *Public Assistance Alternative Procedures Permanent Work Guide* for the USVI to provide guidance on the implementation of the program in the territory. FEMA and USVI officials stated that a section of the Bipartisan Budget Act of 2018 and the flexibilities provided by the program itself influenced the USVI's decision to transition to using the alternative procedures.

First, Section 20601 of the Bipartisan Budget Act of 2018 authorized FEMA, when using the Public Assistance alternative procedures, to

---

[24] According to FEMA officials, these data represent projects that have already been submitted to FEMA for federal review and projects that FEMA expects to receive in the future. Further, the $779.4 million estimate represents the total cost of completing all projects, not the total federal share FEMA will be responsible for obligating.

[25] In addition to the $779.4 million estimate above, FEMA also projects an additional $7.4 million for management costs under category Z.

[26] The Sandy Recovery Improvement Act of 2013 authorized the use of alternative procedures in administering the Public Assistance program, thereby providing new flexibilities to FEMA, states, and local governments using funds from this program. See 42 U.S.C. § 5189f. The stated goals of the alternative procedures are to reduce the costs to the federal government, increase flexibility in the administration of the Public Assistance program, expedite the provision of assistance under the program, and provide financial incentives for recipients of the program for the timely and cost-effective completion of projects.

provide assistance to fund the replacement or restoration of disaster-damaged infrastructure that provide critical services to industry standards without regard to pre-disaster condition.[27] FEMA and USVI officials told us that the territory therefore has a valuable opportunity to use the alternative procedures to repair and rebuild its critical services infrastructure—including the USVI's education system, electrical grid, and emergency medical care system, among others—so it is in a better condition than it was prior to the 2017 hurricanes.[28] Second, USVI officials stated that under the standard Public Assistance program currently being used in the USVI, the territorial government is responsible for providing the initial funding to reimburse subrecipients for completed work prior to drawing down funds from the account holding the FEMA- obligated amounts of money for each project. They explained that because of the financial liquidity challenges facing the territory, this process was problematic and required USVI officials to prioritize projects based on the availability of the territory's funding. USVI officials stated that the Public Assistance alternative procedures will help to address this challenge by providing the territory with more flexibility regarding when and how to fund projects. For example, in certain cases, the USVI is able to consolidate permanent work projects approved under the alternative procedures and share obligated funding across these projects. In addition, the USVI is able to use any excess funds for cost-effective hazard mitigation measures or for activities that improve the recipient's or subrecipient's future Public Assistance operations or planning.

As of November 2018, FEMA and USVI officials stated they were working to identify and develop permanent work projects using the Public Assistance alternative procedures and discussing the process for

---

[27] Bipartisan Budget Act of 2018, Pub. L. No.115-123, § 20601(1), 132 Stat. 64 (2018). Critical services include public infrastructure in the following sectors: power, water, sewer, wastewater treatment, communications, education, and emergency medical care. See 42 U.S.C. § 5172(a)(3)(B). Section 20601 applies only to assistance provided through the Public Assistance alternative procedures program for the duration of the recovery for the major disasters declared in Puerto Rico and the USVI following hurricanes Irma and Maria.

[28] Under the standard Public Assistance program, FEMA generally estimates the eligible cost of repairing, restoring, reconstructing, or replacing a public facility or private nonprofit facility on the basis of the design of the facility as it existed immediately before the major disaster. See 42 U.S.C. § 5172(e).

developing the fixed-cost estimate for each project. Specifically, unlike in the standard Public Assistance program where FEMA will fund the actual cost of a project, the Public Assistance alternative procedures use a fixed-cost estimate which is agreed to prior to obligation and the USVI will be financially responsible for any actual costs that exceed this amount.

Given the USVI's difficult fiscal situation, FEMA and USVI officials stated that ensuring these fixed-cost estimates are as accurate as possible will be critical. However, FEMA officials also noted that if FEMA and the territory cannot come to an agreement on a fixed-cost estimate for any given project, the USVI does have the option to move forward through the standard Public Assistance program. According to FEMA's *Public Assistance Alternative Procedures Permanent Work Guide* for the USVI, all cost estimates for projects using these procedures must be finalized by March 2020.[29] We will continue to monitor the USVI's plans for using the alternative procedures as part of our broader work assessing disaster recovery efforts in the USVI and will issue a follow-on report later this year.

AGENCY COMMENTS

We provided a draft of this chapter to DHS and the USVI government. We requested comments from DHS and the USVI government, but none were provided. DHS did provide technical comments, which we incorporated as appropriate.

We are sending copies of this chapter to the appropriate congressional committees, the Secretary of Homeland Security, the Administrator of FEMA, the USVI government, and other interested parties.

Chris P. Currie
Director, Homeland Security and Justice

---

[29] According to FEMA officials, FEMA and the USVI are reviewing projects that have already been obligated funding under FEMA's standard Public Assistance program for inclusion in the alternative procedures program.

List of Requesters

The Honorable Michael Enzi
Chairman
Committee on the Budget
United States Senate

The Honorable Ron Johnson
Chairman

The Honorable Gary C. Peters
Ranking Member
Committee on Homeland Security and Governmental Affairs
United States Senate

The Honorable Rand Paul, MD
Chairman
Subcommittee on Federal Spending Oversight and
Emergency Management
Committee on Homeland Security and Governmental Affairs
United States Senate

The Honorable Bennie G. Thompson
Chairman
Committee on Homeland Security
House of Representatives

The Honorable Elijah Cummings
Chairman
Committee on Oversight and Reform
House of Representatives

The Honorable Peter DeFazio
Chairman

The Honorable Sam Graves
Ranking Member
Committee on Transportation and Infrastructure
House of Representatives

The Honorable Maxine Waters
Chairwoman
Committee on Financial Services
House of Representatives

The Honorable Nydia Velázquez
Chairwoman
Committee on Small Business
House of Representatives

The Honorable Al Green
Chairman
Subcommittee on Oversight and Investigations
Committee on Financial Services
House of Representatives

The Honorable Sean Duffy
Ranking Member
Subcommittee on Housing, Community Development and Insurance
Committee on Financial Services
House of Representatives

The Honorable Emanuel Cleaver, II
House of Representatives

The Honorable Michael McCaul
House of Representatives

The Honorable Gary Palmer
House of Representatives

The Honorable Ann Wagner
House of Representatives

In: Hurricanes, Wildfires and Flooding ISBN: 978-1-53616-362-9
Editor: Lydie Yohan

Chapter 7

MAJOR DISASTER ASSISTANCE FROM THE DRF: U.S. VIRGIN ISLANDS*

Bruce R. Lindsay

OVERVIEW

The Disaster Relief Fund (DRF) is managed by the Federal Emergency Management Agency (FEMA) and is the primary source of federal assistance following a major disaster declaration. The authority to declare a major disaster is provided to the President under the Robert T. Stafford Emergency Relief and Disaster Assistance Act (P.L. 93-288, as amended; Stafford Act).

The information below includes only DRF funding provided to the U.S. Virgin Islands as a result of assistance associated with a major disaster declaration declared between FY2000 and FY2016. This information does not include other assistance provided through the DRF or disaster assistance provided by other federal sources, such as Small

* This is an edited, reformatted and augmented version of Congressional Research Service Publication No. IF10756, dated October 16, 2017.

Business Administration disaster loans. In FY2017, two major disasters were declared for the U.S. Virgin Islands. Those major disaster declarations are not included in the analysis below.

FEMA administers disaster relief through regional offices. The U.S. Virgin Islands are part of Region II, headquartered in New York, NY. The U.S. Virgin Islands are considered a state under Section 102 of the Stafford Act (42 U.S.C. 5122) for the purposes of the act.

MAJOR DISASTER DECLARATIONS: FY2000-FY2016

A total of seven major disaster declarations were made for the U.S. Virgin Islands between FY2000 and FY2016. During that time, there were no requests for major disaster declarations that were denied. The approved declarations led to $45 million in actual and projected federal obligations from the DRF. Actual and projected obligations are used throughout this In Focus to account for the total amount of federal assistance ultimately obligated. This funding does not include the assistance that was provided directly by the territory, either as a cost-share for federal assistance or through the territory's own authorities and programs.

There was a high level of variation in the amount of actual and projected funding obligated for major disasters each year, with more than $14 million in actual and projected obligations for disasters in FY2000 alone. *Figure 1* displays the actual and projected obligations to the U.S. Virgin Islands each fiscal year. The highest number of declarations in a single year was two, which occurred in FY2011.

MOST COSTLY DECLARATIONS

The declaration associated with the largest actual and projected obligations from the DRF to the U.S. Virgin Islands was for Hurricane

Lenny in FY2000. This declaration resulted in more than $14 million in actual and projected obligations to the U.S. Virgin Islands from the DRF.

The next two most costly declarations were for Hurricane Omar in FY2009 and severe storms, flooding, rockslides, and mudslides in FY2011.

Local Impacts

From FY2000 to FY2016, three local or territorial governments in the U.S. Virgin Islands received aid under major disaster declarations. The island of St. Croix received aid for the highest number of declarations during this period, with seven.

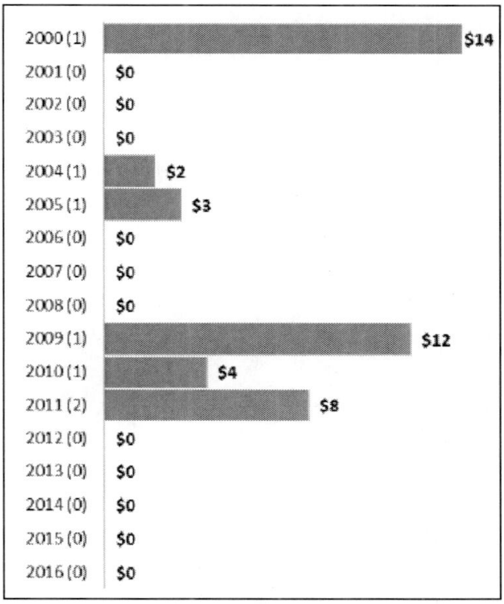

Source: CRS analysis of FEMA DRF obligation data as of Aug. 2017.

Notes: Actual and projected obligations are recorded in the fiscal year in which the major disaster was declared. Only obligations from the FEMA DRF are included.

Figure 1. DRF Actual and Projected Obligations for Declared Major Disasters for the U.S. Virgin Islands, by Fiscal Year. Figures are in millions of dollars, truncated. The number of declarations per year is in parentheses.

Types of Assistance Provided for Major Disasters

A major disaster declaration can authorize funding for different purposes, depending on the needs of the territory. These purposes include:

- *Public Assistance*, which is used by territorial or local governments, or certain private nonprofit organizations to provide emergency protective services, conduct debris removal operations, and repair or replace damaged public infrastructure;
- *Individual Assistance,* which provides direct aid to impacted households;
- *Hazard Mitigation Assistance*, which funds mitigation and resiliency projects and programs, typically across the entire territory;
- *FEMA administrative costs* associated with each disaster declaration; and
- *Mission Assignments*, which tasks and reimburses other federal entities that provide direct disaster assistance.

The decision concerning which types of assistance to provide is made either when the major disaster is declared or when the declaration is amended. For many major disasters, all of the assistance types outlined above are authorized. For others, some assistance types are not authorized. *Figure 2* compares the actual and projected obligations for different types of assistance provided to the U.S. Virgin Islands as a result of a major disaster declaration from FY2000 to FY2016.

Other FEMA Assistance Funded Through the DRF

In addition to the major disaster assistance described above, there are other forms of assistance that are funded through the DRF. Funding associated with these types of assistance typically results in lower obligation levels than assistance through major disaster declarations, although there is significant variation across incidents.

- *Emergency Declarations* are often made when a threat is recognized to assist territorial preparation efforts.
- *Fire Management Assistance Grants (FMAGs)* provide aid for the control, management, and mitigation of fires.

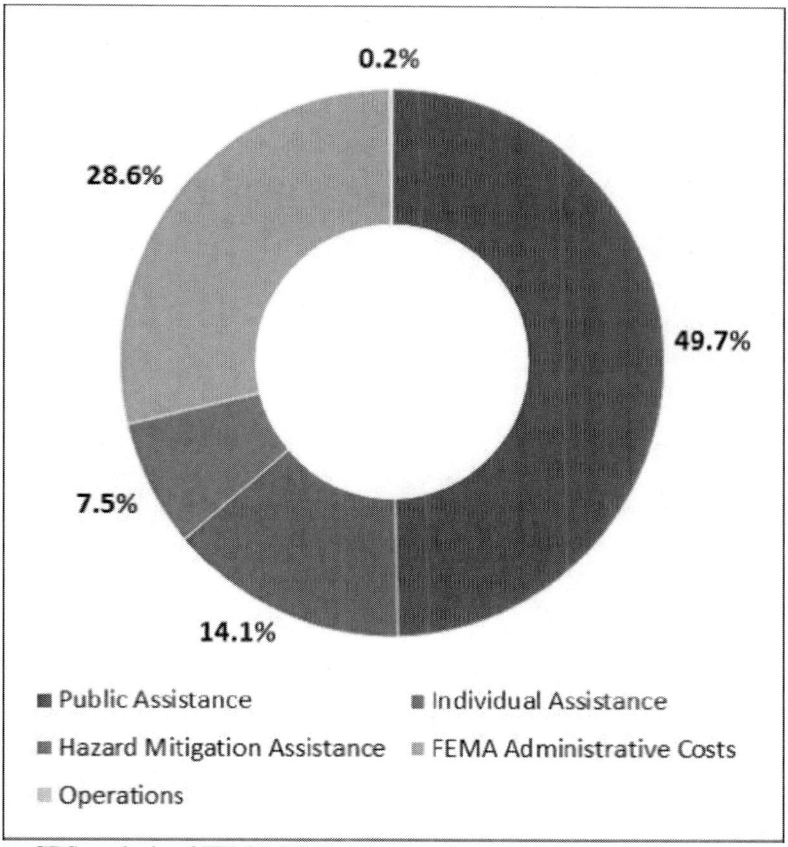

Source: CRS analysis of FEMA DRF obligation data as of Aug. 2017.

Note: The specific assistance within each of the identified types may have changed in the period FY2000 to FY2016. "Operations" is a category provided by FEMA that includes three types of Mission Assignment and Urban Search and Rescue.

Figure 2. DRF Assistance for the U.S. Virgin Islands, by Type (FY2000-FY2016).

FEMA Assistance Not Funded Through the DRF: National Flood Insurance Program

Floods represent a majority of all major disaster declarations nationwide. One of the primary sources of assistance for flooding events is the National Flood Insurance Program (NFIP), which is not funded through the DRF.

In: Hurricanes, Wildfires and Flooding
Editor: Lydie Yohan
ISBN: 978-1-53616-362-9

Chapter 8

MAJOR DISASTER ASSISTANCE FROM THE DRF: FLORIDA*

Bruce R. Lindsay

OVERVIEW

The Disaster Relief Fund (DRF) is managed by the Federal Emergency Management Agency (FEMA) and is the primary source of federal assistance following a major disaster declaration. The authority to declare a major disaster is provided to the President under the Robert T. Stafford Emergency Relief and Disaster Assistance Act (P.L. 93-288).

The information below includes only DRF funding provided to Florida as a result of assistance associated with a major disaster declaration. This information does not include other assistance provided through the DRF or disaster assistance provided by other federal sources, such as Small Business Administration disaster loans.

---

* This is an edited, reformatted and augmented version of Congressional Research Service Publication No. IF10067, Updated March 6, 2017.

FEMA administers disaster relief through regional offices around the country. Florida is part of Region IV, headquartered in Atlanta, GA.

MAJOR DISASTER DECLARATIONS: FY2000- FY2015

A total of 25 major disaster declarations were made in the state of Florida between FY2000 and FY2015. During that time, there were also nine requests for major disaster declarations that were denied. The approved declarations led to $10.8 billion in actual and projected federal obligations from the DRF. Actual and projected obligations are used throughout this In Focus to account for the total amount of federal assistance ultimately obligated. This funding does not include the assistance that was provided directly by the state, either as a cost-share for federal assistance or through the state's own authorities and programs.

There was a high level of variation in the amount of actual and projected funding obligated for major disasters each year, with more than $6.24 billion in actual and projected obligations for disasters in FY2004 alone. *Figure 1* displays the actual and projected obligations to Florida each fiscal year. The highest number of declarations in a single year was five, which occurred in FY2001.

MOST COSTLY DECLARATIONS

The declaration associated with the largest actual and projected obligations from the DRF to the state of Florida was for Hurricane Wilma in FY2006. This declaration resulted in more than $2.57 billion in actual and projected obligations to Florida from the DRF.

Other costly incidents include Hurricanes Charley, Francis, Ivan, and Jeanne in FY2004.

LOCAL IMPACTS

From FY2000 to FY2015, 68 county governments in Florida received aid under major disaster declarations. Collier County received aid for the highest number of declarations during this period, with 13.

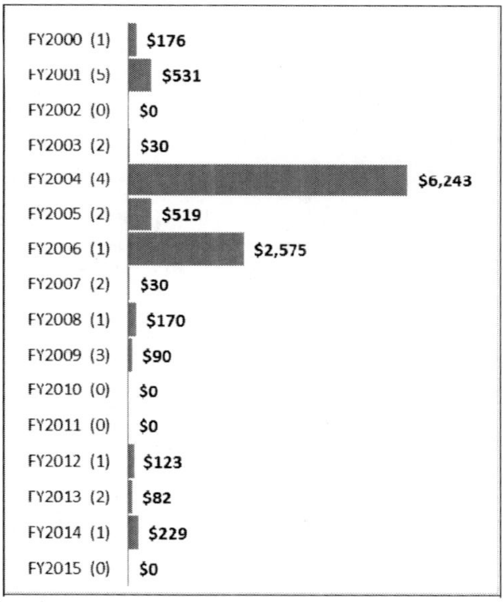

Source: CRS analysis of FEMA DRF obligations data as of December 2015.

Notes: Actual and projected obligations are recorded in the fiscal year in which the major disaster was declared. Only obligations from the FEMA DRF are included. Please consult the "Additional Resources" section for information on other federal assistance programs.

Figures are in millions of dollars, truncated. The number of declarations per year is in parentheses.

Figure 1. DRF Actual and Projected Obligations for Declared Major Disasters in Florida, by Fiscal Year.

Types of Assistance Provided for Major Disasters

A major disaster declaration can authorize funding for different purposes, depending on the needs of the state. These purposes include:

- Public Assistance, which is used by tribal, state, or local governments, or certain private nonprofit organizations to provide emergency protective services, conduct debris removal operations, and repair or replace damaged public infrastructure; to impacted households;
- Hazard Mitigation Assistance, which funds mitigation and resiliency projects and programs, typically across the entire state;
- FEMA administrative costs associated with each disaster declaration; and
- Mission Assignment, which tasks and reimburses other federal entities that provide direct disaster assistance.

The decision concerning which types of assistance to provide is made either when the major disaster is declared or when the declaration is amended. For many major disasters, all of the assistance types outlined above are authorized. For others, some assistance types are not authorized. *Figure 2* compares the actual and projected obligations for different types of assistance provided to Florida as a result of a major disaster declaration from FY2000 to FY2015.

Other FEMA Assistance Funded Through the DRF

In addition to the major disaster assistance described above, there are other forms of assistance that are funded through the DRF. These include assistance associated with Emergency Declarations and with Fire Management Assistance Grants. The funding associated with these types of assistance typically results in lower obligation levels than assistance provided as a result of major disaster declarations, although there is significant variation across incidents. assistance provided as a result of major disaster declarations, although there is significant variation across incidents.

- Emergency Declarations are often made at the time a threat is recognized in order to assist tribal, state, and local efforts prior to an incident.
- Fire Management Assistance Grants (FMAGs) provide aid for the control, management, and mitigation of fires.

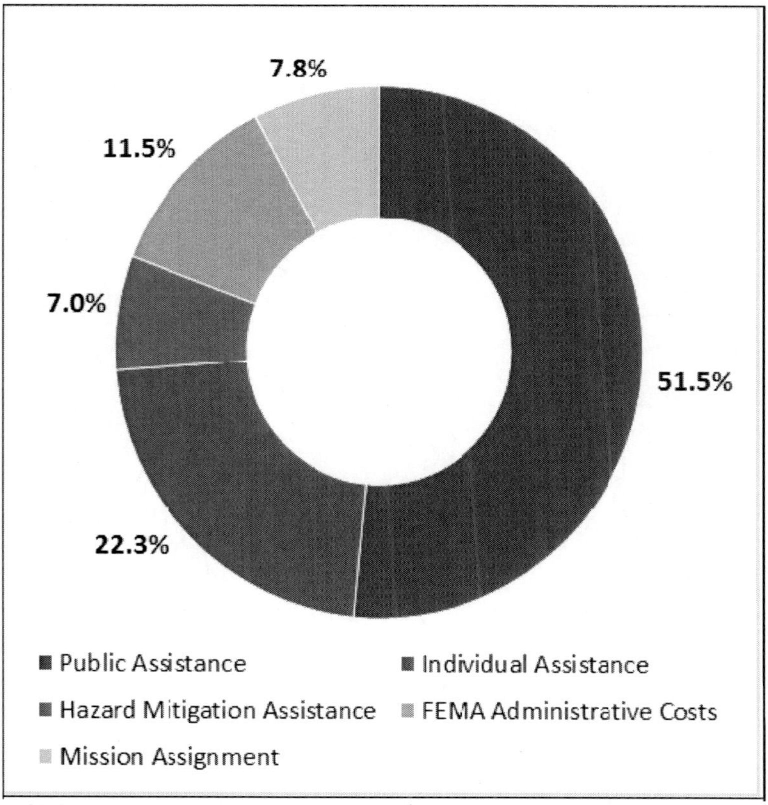

Source: CRS analysis of FEMA DRF obligation data as of December 2015.

Note: The specific assistance within each of the identified types may have changed in the period FY2000 to FY2015.

Figure 2. DRF Assistance for Florida, by Type (FY2000-FY2015).

FEMA Assistance Not Funded Through the DRF: National Flood Insurance Program

Floods represent a majority of all major disaster declarations nationwide. One of the primary sources of assistance for flooding events is the National Flood Insurance Program (NFIP), which is not funded through the DRF.

In: Hurricanes, Wildfires and Flooding ISBN: 978-1-53616-362-9
Editor: Lydie Yohan

Chapter 9

2019 MIDWEST FLOODING: FEMA AND OTHER FEDERAL PROGRAMS AND RESOURCES*

Elizabeth M. Webster, Bruce R. Lindsay, Diane P. Horn, William L. Painter and Natalie Keegan

INTRODUCTION

This Insight provides a brief overview of the major disaster declaration process and federal assistance programs potentially available to those affected by the current flooding in the Midwest.

As authorized under the Robert T. Stafford Disaster Relief and Emergency Assistance Act (P.L. 93-288, as amended; 42 U.S.C. §§5121 et seq.), the President may issue a major disaster declaration due to

---

* This is an edited, reformatted and augmented version of a Congressional Research Service publication IN11084, prepared for Members and Committees of Congress dated March 26, 2019.

catastrophic flooding. This allows for a broad range of federal assistance programs to be made available to state and local governments, private nonprofit organizations, and individuals through the Federal Emergency Management Agency (FEMA) and other federal agencies.

According to the National Oceanic and Atmospheric Administration (NOAA), the record flooding in Nebraska, Iowa, and Minnesota is the result of "rapid snow melt combined with heavy spring rain and late season snowfall in areas where soil moisture is high."

On March 21, 2019, President Donald J. Trump issued a major disaster declaration for the State of Nebraska for the severe winter storm, straight-line winds, and flooding that began on March 9, 2019. On March 23, 2019, President Trump issued a major disaster declaration for the State of Iowa for the severe storms and flooding that began on March 12, 2019.

Other states affected by flooding have declared states of emergency, including Missouri, Kansas, Wisconsin, Ohio, and South Dakota as well as tribal nations in Nebraska, Iowa, and South Dakota. Additionally, February rainfall and severe storms that resulted in flooding in Mississippi and Tennessee resulted in their governors' submission of requests for major disaster declarations on March 21 and March 22, 2019, respectively.

NOAA's spring outlook states that additional rain and melting snow "will prolong and expand flooding," and that as water flows downstream, the threat of flooding will increase and become more widespread.

Addressing Sheltering and Housing Needs

Many structures have been damaged or destroyed by the flooding, including estimates of more than 2,000 homes in Nebraska, according to Governor Pete Ricketts, and more than 1,000 structures—including homes—in Iowa, according to Governor Kim Reynolds. FEMA, working with the state government, is scheduled to begin the damage assessment process in Nebraska the week of March 25, 2019. Estimates may change as damage assessments are conducted.

The American Red Cross is operating shelters in states affected by flooding, including Nebraska, Iowa, Missouri, Wisconsin, Ohio, Mississippi, and Illinois, and emergency protective measures (e.g., sheltering assistance) and Individual Assistance (IA) are authorized for eligible individuals and households residing in designated areas in Nebraska and Iowa, pursuant to the presidential declarations. Additionally, the Small Business Administration (SBA) may provide low-interest loans to homeowners/renters to cover residential losses that are not fully compensated by insurance.

Types of FEMA Assistance Provided for Major Disasters

FEMA provides three major categories of assistance for major disasters:

- *Public Assistance* (PA) provides grants to tribal, state, and local governments, and certain private nonprofit organizations for emergency protective measures, debris removal operations, and repair or replacement of damaged public infrastructure.
- *Individual Assistance* (IA) provides direct aid to affected individuals and households, and can take the form of housing assistance, other needs assistance, crisis counseling, case management services, legal services, and disaster unemployment assistance.
- *Hazard Mitigation Assistance* (HMA) funds mitigation and resiliency projects and programs, typically across the entire state.

The forms of assistance authorized by a major disaster declaration may vary by the designated areas, per the declaration (subject to amendment). The President can also amend major disaster declarations to decrease the state cost-share requirements for some PA grants.

National Flood Insurance Program

The National Flood Insurance Program (NFIP) is the primary source of flood insurance coverage for residential properties. Homeowners and tenants with NFIP flood insurance can make flood damage claims using the normal NFIP claims process. FEMA may institute a claims process specific to the Midwest floods at a later date; for example, FEMA extended the proof of loss deadline for Hurricane Florence (2018) from 60 days to 1 year.

As of January 1, 2019, the NFIP had $5.317 billion available ($4.496 billion in the National Flood Insurance Fund and $822 million in the reserve fund), as well as $9.9 billion of borrowing authority from the Treasury and up to $1.96 billion of reinsurance for a single flood event with losses over $4 billion.

Federal Funding for Disaster Response and Recovery

FEMA's Disaster Relief Fund (DRF) is the primary source of federal government resources for response and recovery activities. It is often used as an indicator of the overall availability of federal resources for response and recovery.

As a result of more than $12 billion in annual appropriations provided in P.L. 116-6, Division A, and carryover balances from prior-year appropriations, as of February 28, 2019, the DRF had nearly $31 billion in unobligated budget authority. These funds do not expire at the end of the fiscal year, and are to remain available until expended.

Several other disaster relief programs rely on supplemental appropriations to support their operations. These include the SBA Disaster Loan Program and the Department of Housing and Urban Development's Community Development Fund. In October 2018, the latter received a $1.68 billion supplemental appropriation in Division I of P.L. 115-254 for grants to states for disaster relief and recovery activities "in the most impacted and distressed areas resulting from a major disaster declared in 2018." Prior to this supplemental appropriation, the most recent supplemental appropriations for disaster response and recovery were provided in P.L. 115-123, enacted in February 2018.

ADDITIONAL FLOODING INCIDENT RESOURCES

- Citizens:
  - Ready.gov Floods
  - Red Cross Flood Safety Checklist
  - National Weather Service Social Media: Flood Safety
- FEMA Programs:
  - Public Assistance Fact Sheet
  - Individual Assistance Fact Sheet
  - Hazard Mitigation Grant Program Overview
  - National Flood Insurance Program Overview

In: Hurricanes, Wildfires and Flooding
Editor: Lydie Yohan
ISBN: 978-1-53616-362-9

Chapter 10

2017 Disaster Contracting: Action Needed to Better Ensure More Effective Use and Management of Advance Contracts*

United States Government Accountability Office

Abbreviations

DOD	Department of Defense
DHS	Department of Homeland Security
FAR	Federal Acquisition Regulation
FEMA	Federal Emergency Management Agency
FPDS-NG	Federal Procurement Data System-Next Generation

---

* This is an edited, reformatted and augmented version of United States Government Accountability Office; Report to Congressional Requesters, Publication No. GAO-19-93, dated December 2018.

IASC	Individual Assistance Support Contract
IA-TAC	Individual Assistance-Technical Assistance Contract
LOGHOUSE	Logistics Housing Operations Unit Installation, Maintenance, and Deactivation Contracts
MAPS	Master Acquisition Planning Schedule
PA-TAC	Public Assistance-Technical Assistance Contract
PKEMRA	Post-Katrina Emergency Management Reform Act of 2006
USACE	U.S. Army Corps of Engineers

Why GAO Did This Study

Following Hurricane Katrina, Congress required FEMA to establish advance contracts for goods and services to enable the government to quickly and effectively mobilize resources in the aftermath of a disaster, like those that affected the United States in 2017.

GAO was asked to review the federal government's response to the three 2017 hurricanes and California wildfires. This chapter assesses, among other things, (1) FEMA and USACE's use of advance contracts, (2) FEMA's planning and reporting of selected advance contracts, and (3) challenges, if any, with FEMA's use of these contracts.

GAO analyzed data from the Federal Procurement Data System-Next Generation through May 31, 2018; selected a non-generalizable sample of 14 FEMA and USACE advance contracts that were competed and obligated over $50 million, or non-competed and obligated over $10 million, in response to the 2017 disasters; and interviewed FEMA and USACE officials.

WHAT GAO RECOMMENDS

GAO is making nine recommendations to FEMA, including that it update its strategy and guidance to clarify the use of advance contracts, improve the timeliness of its acquisition planning activities, revise its methodology for reporting disaster contracting actions to congressional committees, and provide more consistent guidance and information to contracting officers to coordinate with and encourage states and localities to establish advance contracts. FEMA concurred with our recommendations.

WHAT GAO FOUND

In response to Hurricanes Harvey, Irma, and Maria, as well as the 2017 California wildfires, the Federal Emergency Management Agency (FEMA) and U.S. Army Corps of Engineers (USACE) relied heavily on advance contracts. As of May 31, 2018, FEMA and USACE obligated about $4.5 billion for various goods and services through these contracts, see figure below.

GAO found limitations in FEMA's use of some advance contracts that provided critical goods and services to survivors, including

- an outdated strategy and unclear guidance on how contracting officers should use advance contracts during a disaster, and
- challenges performing acquisition planning.

FEMA also did not always provide complete information in its reports to congressional committees. Specifically, GAO found 29 advance contract actions that were not included in recent reports due to shortcomings in FEMA's reporting methodology, limiting visibility into its disaster contract spending.

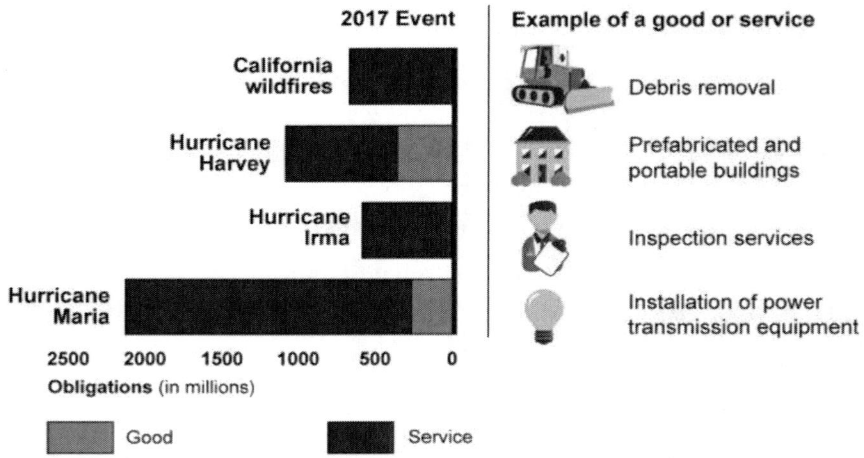

Source: GAO analysis of Federal Procurement Data System-Next Generation data. | GAO-19-93.

Note: Obligations are nominal 2017 and 2018 dollars.

Advance Contract Obligations and Examples of Goods and Services the Federal Emergency Management Agency and U.S. Army Corps of Engineers Used to Respond to the 2017 Disasters.

FEMA identified challenges with advance contracts in 2017, including federal coordination with states and localities on their use. FEMA is required to coordinate with states and localities and encourage them to establish their own advance contracts with vendors. However, GAO found inconsistencies in that coordination and the information FEMA uses to coordinate with states and localities on advance contracts. Without consistent information and coordination with FEMA, states and localities may not have the tools needed to establish their own advance contracts for critical goods and services and quickly respond to future disasters.

December 6, 2018

Congressional Requesters

The Post-Katrina Emergency Management Reform Act of 2006 (PKEMRA) required the Federal Emergency Management Agency (FEMA) to establish advance contracts—those that are established prior to disasters and that are typically needed to quickly provide life-sustaining goods and services in the immediate aftermath of disasters.[1] FEMA and the U.S. Army Corps of Engineers (USACE)—components of the Department of Homeland Security (DHS) and the Department of Defense (DOD), respectively—used advance contracts to procure goods and services provided in response to the catastrophic disasters the United States experienced in 2017: Hurricanes Harvey, Irma, and Maria; and the California wildfires.[2]

Contracts play a key role in immediate disaster response and longer-term community recovery. In February 2018, we found that early estimates for Hurricanes Harvey, Irma, and Maria placed them among the costliest hurricanes in terms of federal contract obligations since 2005, when tracking this information by hurricane began.[3] We also found that DHS and DOD accounted for about 97 percent of those obligations. In September 2015, we found issues related to the use of advance contracts, specifically that FEMA's contracting officers had limited awareness about advance contract requirements and how to coordinate advance contracting efforts with state and local governments.[4] We made several recommendations, including that FEMA provide guidance to contracting officers to make them aware of current information on the availability and use of advance contracts and the need to conduct outreach to state and local governments to support their use of advance contracts. FEMA agreed with our recommendations and has taken action to address them.

You asked us to review the federal government's contracting efforts for preparedness, response, and recovery efforts related to the three 2017 hurricanes and California wildfires. This chapter specifically addresses the

---

[1] Pub. L. No. 109-295, § 691 (codified at 6 U.S.C. § 791).

[2] Although PKEMRA does not apply to USACE, the agency awards advance contracts as a preparedness measure as part of its Advance Contracting Initiative, a program developed and implemented specifically for emergency and disaster scenarios. For the purposes of this chapter, we rely on how FEMA and USACE define advance contracts. For FEMA, this includes indefinite delivery contracts and blanket purchase agreements, including those under the Federal Supply Schedule, as well as interagency agreements, and interagency reimbursable work agreements. For USACE this includes indefinite delivery contracts. Obligations against interagency agreements are not included in this chapter.

[3] GAO, *2017 Disaster Contracting: Observations on Federal Contracting for Response and Recovery Efforts*, GAO-18-335 (Washington D.C.: Feb. 28, 2018).

[4] GAO, *Disaster Contracting: FEMA Needs to Cohesively Manage Its Workforce and Fully Address Post-Katrina Reforms*, GAO-15-783 (Washington, D.C.: Sept. 29, 2015).

use of advance contracts, assessing the extent to which (1) FEMA and USACE used advance contracts, (2) the planning, management, and reporting of selected FEMA and USACE advance contracts met certain contracting requirements, and (3) FEMA and USACE identified any lessons learned and challenges with their use of these contracts. We also have an ongoing review on post-disaster contracting that is expected to be completed in early 2019.

To identify the extent to which FEMA and USACE used advance contracts, we reviewed data on contract obligations for the 2017 disasters from the Federal Procurement Data System-Next Generation (FPDS-NG) through May 31, 2018.[5] Data on obligations for the California wildfires are limited to contracts that FEMA and USACE identified as being used to respond to the events because some of this information was not able to be identified in FPDS-NG. To determine which obligations were made through the use of advance contracts, we reviewed documentation provided by FEMA and USACE identifying the advance contracts they have in place and that were used in support of the 2017 disasters.[6] We assessed the reliability of FPDS-NG data by reviewing existing information about the FPDS-NG system and the data it collects— specifically, the data dictionary and data validation rules—and performing electronic testing. We determined the FPDS-NG data were sufficiently reliable for the purposes of this chapter. We also reviewed FEMA contracting policies and guidance, to identify available guidance on the use and intent of advance

[5] For the purposes of this chapter, contract obligations include obligations against what the General Services Administration's Federal Procurement Data System-Next Generation (FPDS-NG) categorizes as definitive vehicles (definitive contracts and purchase orders that have a defined scope of work that do not allow for individual orders under them), and against what FPDS-NG categorizes as indefinite delivery vehicles (orders under the Federal Supply Schedule, orders/calls under blanket purchase agreements, orders under basic ordering agreements, orders under government-wide acquisition contracts, and orders under other indefinite delivery vehicles, such as indefinite delivery, indefinite quantity contracts).

[6] DHS and DOD exercised the use of the special emergency procurement authorities within 41 U.S.C § 1903, as amended by § 816 of the National Defense Authorization Act for Fiscal Year 2017, Pub. L. No. 114-328 (and as implemented by DHS FAR Class Deviation 17-02 and DOD FAR Class Deviation 2017-O0007, respectively) to increase the micro-purchase threshold to $20,000 for procurements in support of these major disaster responses. As a result, contract obligations for hurricanes reported in FPDS-NG may only include obligations over that amount.

contracts, and federal internal control standards. We identified examples of goods that FEMA had advance contracts in place for but experienced challenges using in response to the 2017 disasters, reviewed advance and post-disaster contract documentation and FPDS-NG data related to these examples, and interviewed contracting officials involved in the award and use of the contracts in 2017.

To assess the extent to which the planning, management, and reporting of advance contracts used in response to the three hurricanes and California wildfires in 2017 met selected applicable contracting requirements, we reviewed relevant documentation, including PKEMRA, the Federal Acquisition Regulation (FAR), DHS, FEMA, and USACE contracting policies. We identified a non-generalizable sample of advance contracts to serve as case studies based on obligation data from FPDSNG as of March 31, 2018. We analyzed the data to identify 10 competed and 4 non-competed contracts—valued at above $50 million and above $10 million, respectively—including 10 contracts from FEMA and 4 from USACE.[7] For additional details on the contracts we selected, see appendix I. Findings based on information collected from the 14 contracts cannot be generalized to all advance contracts. To review our selected FEMA and USACE advance contracts, we developed a data collection instrument to gather selected contract information, such as period of performance, contract type, estimated contract value, and the presence of key contract documents, among others. To assess planning we reviewed our selected advance contracts, and determined that six of FEMA's contracts met GAO's definition of a bridge contract. We interviewed FEMA officials associated with these contracts on acquisition planning efforts and factors that affected their ability to award new contracts. To assess FEMA and USACE's management of selected advance contracts, we reviewed information gathered from our data collection instrument to confirm the contract files contained required acquisition documents, such as acquisition

[7] For the purposes of selecting our case studies, competitive contracts included contracts and orders coded in FPDS-NG as "full and open competition," "full and open after exclusion of sources," and "competed under simplified acquisition procedures." Noncompetitive contracts included contracts and orders coded in FPDS-NG as "not competed," "not available for competition," and "not competed under simplified acquisition procedures."

strategies and contract modifications, which typically provide the history of a contract file, and interviewed officials at FEMA and USACE headquarters on their record keeping policies, practices, and challenges. To assess the reporting of selected advance contracts, we compared advance contract action data identified in FPDS-NG in the fourth quarter of fiscal year 2017 and first quarter of fiscal year 2018 to data reported for the same time period in FEMA's mandated reports to congressional committees on disaster contracting to identify any unreported actions. We interviewed FEMA officials to discuss the methodology and data sources for the congressional committee reports and any limitations to the accuracy of the data reported.

To assess what lessons learned and challenges FEMA and USACE identified with the use of advance contracts in 2017, we interviewed FEMA and USACE officials on what they identified as lessons learned and challenges specific to advance contracts, and any recommendations or actions planned by the agencies to address them. We reviewed PKEMRA advance contract requirements, including requirements that FEMA coordinate with state and local governments on the use of advance contracts, FEMA and USACE documentation on the use of advance contracts, after-action reports from 2017 and prior years, and federal internal control standards for information and communications. To identify challenges related to coordination with state and local officials on the use of advance contracts, we reviewed after action reports from 2017, and interviewed FEMA and USACE regional staff and state and local government officials on advance contracting efforts. We also analyzed information on available advance contracts from FEMA's June 2018 Prepositioned Contract List and FEMA's May 2018 training documentation identifying advance contracts to identify any differences in the information available to FEMA contracting officers, and their state and local contracting counterparts.[8] Appendix I provides more information about our overall scope and methodology.

---

[8] Prepositioned contracts are the same as advance contracts. For the purposes of this chapter, we use the term advance contracts and refer to FEMA's Prepositioned Contract List as the advance contract list.

We conducted this performance audit from March 2018 to December 2018 in accordance with generally accepted government auditing standards. Those standards require that we plan and perform the audit to obtain sufficient, appropriate evidence to provide a reasonable basis for our findings and conclusions based on our audit objectives. We believe that the evidence obtained provides a reasonable basis for our findings and conclusions based on our audit objectives.

BACKGROUND

The 2017 Hurricanes and California Wildfires

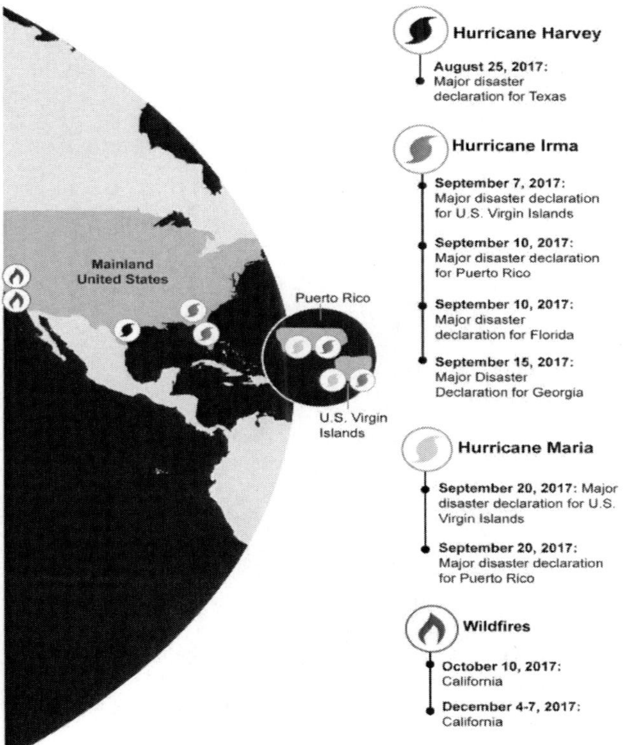

Source: GAO analysis of Federal Emergency Management Agency data. | GAO-19-93.

Figure 1. Timeline of Major 2017 Hurricanes and California Wildfires.

In 2017, three major hurricanes made landfall in the United States and historic wildfires struck California. According to FEMA, the 2017 hurricanes and wildfires collectively affected 47 million people—nearly 15 percent of the nation's population. See Figure 1 for a timeline of these major disasters.[9]

Overview of Federal Disaster Response and Recovery

When disasters hit, state and local entities are typically responsible for disaster response efforts. The Robert T. Stafford Disaster Relief and Emergency Assistance Act established a process by which a state may request a presidential disaster declaration to obtain federal assistance.[10] According to the DHS National Response Framework—a guide to how the federal government, states and localities, and other public and private sector institutions should respond to disasters and emergencies—the Secretary of Homeland Security is responsible for ensuring that federal preparedness actions are coordinated to prevent gaps in the federal government's efforts to respond to all major disasters, among other emergencies.[11] The framework also designates FEMA to lead the coordination of the federal disaster response efforts across federal agencies.

The National Response Framework identifies 14 emergency support functions that serve as the federal government's primary coordinating

---

[9] A major disaster is any natural catastrophe (including any hurricane, tornado, storm, high water, wind driven water, tidal wave, tsunami, earthquake, volcanic eruption, landslide, mudslide, snowstorm, or drought), or, regardless of cause, any fire, flood, or explosion, in any part of the United States, which the President determines causes damage of sufficient severity and magnitude to warrant major disaster assistance to supplement the efforts and available resources of states, local governments, and disaster relief organizations in alleviating damage, loss, hardship, or suffering. See 42 U.S.C. § 5122(2).

[10] 42 U.S.C. § 5170. The Robert T. Stafford Disaster Relief and Emergency Assistance Act, as amended, permits the President to declare a major disaster after a state's governor or chief executive of an affected Indian tribal government—a governing body of an Indian or Alaska Native tribe, band, nation, pueblo, village, or community that is federally recognized—finds that the disaster is of such severity and magnitude that effective response is beyond his or her own local capacities.

[11] Department of Homeland Security, *National Response Framework* (June 2016).

structure for building, sustaining, and delivering disaster response efforts across more than 30 federal agencies. Each function addresses a specific need—such as communication, transportation, and energy—and designates a federal department or agency as the coordinating agency. For example, the emergency support function for public works and engineering assists DHS by coordinating engineering and construction services, such as temporary roofing or power, and USACE is the primary agency responsible for these functions during disaster response activities.

FEMA coordinates disaster response efforts through mission assignments—work orders that FEMA issues to direct other federal agencies to utilize the authorities and the resources granted to it under federal law. Mission assignments are authorized by the Robert T. Stafford Disaster Relief and Emergency Assistance Act and can consist of federal operations support or direct federal assistance, which includes federal contracts.

FEMA's contracting efforts are supported by its Office of the Chief Procurement Officer and its contracting workforce. While the majority of FEMA's contracting workforce is located in headquarters, contracting officers are also located in each of FEMA's 10 regional offices. See appendix II for the location of FEMA's 10 regional offices as well as the states each one is responsible for coordinating with to address National Response Framework responsibilities.

PKEMRA Requirements and the Use of Advance Contracts

Congress enacted PKEMRA in 2006, which addressed various shortcomings identified in preparation for and response to Hurricane Katrina, which hit the Gulf Coast in 2005 and was one of the largest, most destructive natural disasters in U.S. history. Among the provisions included were requirements for FEMA to identify and establish advance contracts to ensure that goods and services are in place to help FEMA rapidly mobilize resources in immediate response to disasters. Examples of these goods and services are:

- Goods: construction supplies and tarps; food and water; cleaning and hygiene supplies; and power equipment and generators.
- Services: engineering; information technology and communication support; transportation of goods; and housing and lodging assistance.

As of June 2018, FEMA reported having advance contracts in place for 56 different types of goods and services.

Among other contracting requirements, PKEMRA requires FEMA to

- develop a contracting strategy that maximizes the use of advance contracts to the extent practical and cost effective;
- coordinate advance contracts with state and local governments;
- encourage state and local governments to engage in similar pre-planning and contracting; and
- submit quarterly reports to the appropriate committees of Congress on each disaster contract entered into by the agency using non-competitive procedures.[12]

According to FEMA's advance contracting strategy, the agency will maximize the use of advance contracts to the extent they are practical and cost-effective, which will help preclude the need to procure goods and services under unusual and compelling urgency. When disasters strike, contracting officers may use the unusual and compelling urgency exception to full and open competition to support non-competitive contract awards.[13] FEMA's strategy also states that advance contracts will help to ensure that goods and services are in place to help FEMA rapidly mobilize resources in immediate response to disasters.

USACE also has its own advance contracts in place as a preparedness measure. According to USACE officials, they established advance contract

---

[12] While PKEMRA only requires FEMA to report on noncompetitive disaster contract actions, the Joint Explanatory Statement that accompanied PKEMRA recommends that FEMA report quarterly on all contracts issued during a disaster.

[13] Department of Homeland Security, Federal Emergency Management Agency, *Advance Contracting of Goods and Services Report to Congress* (December 2007).

initiatives in 2003, two years prior to Hurricane Katrina, to help facilitate their emergency support function under the National Response Framework—public works and engineering. As of September 2018, USACE reported having advance contracts in place for three services—debris removal, temporary roofing, and temporary power. Appendix III provides details on specific advance contracts established by FEMA and USACE.

According to FEMA documentation, most of its advance contracts are indefinite delivery contracts, which can facilitate the goal of having contracts available if there is a disaster.[14] One type of indefinite delivery contract—an indefinite delivery, indefinite quantity contract—can be awarded to single or multiple vendors and provides for an indefinite quantity, within stated limits, of supplies or services during a fixed period. Under these contracts, the government places orders for individual requirements. These contracts also require the government to order and the contractor to provide at least a stated minimum quantity of supplies and services. Additionally, the contracting officer should also establish a reasonable maximum quantity for the contract based on market research, trends in similar recent contracts, or any other rational basis.[15] Minimum and maximum quantity limits can be stated as the number of units or as dollar values, and may also be referred to by contracting officers as minimum guarantees or contract ceilings, respectively.

As part of its overall acquisition strategy, FEMA officials identified other vehicles aside from its own advance contracts through which they obtain goods and services.

- DHS strategic sourcing vehicles: When a disaster occurs, FEMA contracting officers are first required to use any available DHS strategic sourcing vehicles—a broader, aggregate approach for procuring goods and services—with limited exceptions.

---

[14] Indefinite delivery contracts (definite-quantity contracts, requirements contracts, and indefinite quantity contracts) may be used when the exact times or quantities, or both, of future deliveries are not known at the time of contract award. FAR § 16.501-2.

[15] FAR § 16.504.

- Blanket purchase agreements: FEMA also relies on blanket purchase agreements, such as those established through the General Service Administration Federal Supply Schedule program, to provide some commercial goods and services needed for disaster response.[16]
- Interagency Agreements: FEMA may also leverage interagency agreements, by which it obtains needed supplies or services from another agency by an assisted or direct acquisition.

FEMA and other agencies may also award new contracts to support disaster response efforts following a disaster declaration. According to FEMA officials, these post-disaster contract awards may be required, for example, if advance contracts reach their ceilings, or if goods and services that are not suitable for advance contracts are needed.[17]

FAR Requirements

The FAR requires agencies to perform acquisition planning activities for all acquisitions to ensure that the government meets its needs in the most effective, economical, and timely manner possible.[18] Generally, program and contracting officials share responsibility for the majority of acquisition planning activities, which include the following:

---

[16] The General Services Administration Federal Supply Schedule program provides agencies with a simplified way to purchase commercial goods and services. State and local governments are eligible to use these contracts for certain purchases, including those supporting disaster recovery efforts. Blanket purchase agreements are simplified methods of filling anticipated repetitive needs for supplies or services by establishing "charge accounts" with qualified sources of supply. FAR § 13.303-1. Agencies may establish blanket purchase agreements under the General Services Administration's Schedule program contracts. A blanket purchase agreement is not a contract, therefore the government is not obligated to purchase a minimum quantity or dollar amount, and the contractor is not obligated to perform until it accepts an order under a blanket purchase agreement.

[17] We plan to report on post-disaster contracts in a separate, ongoing review.

[18] FAR § 7.102.

- Pre-Solicitation: The program office identifies a need, and develops key acquisition documents to summarize that need, such as market research, a statement of work defining requirements, cost estimates, and a written acquisition plan.[19] The pre-solicitation process ends when the program office submits these documents, typically referred to as an acquisition package, to the contracting officer to determine what type of contract is appropriate to fulfill the requirements.
- Solicitation: The contracting officer develops a solicitation, in consultation with other agency stakeholders, to request bids or proposals from contractors. The acquisition planning process ends once a solicitation is issued.

Contracting for disaster relief and recovery efforts can also present unique circumstances in which to solicit, award, and administer contracts. Under the FAR, agencies are generally required to use full and open competition when soliciting offers and awarding contracts. However, an agency may award contracts noncompetitively when the need for goods or services is of such unusual and compelling urgency that the federal government faces the risk of serious financial or other type of injury.[20]

---

[19] According to the FAR, contracting agencies should establish criteria and thresholds at which increasingly greater detail and formality in the acquisition planning process is required as the acquisition becomes more complex and costly. Additionally, the FAR states that a written acquisition plan is required for cost reimbursement and other high risk contracts other than firm-fixed-price contracts, although written plans may be required for firm-fixed-price contracts as appropriate. FAR § 7.103(e).

[20] The FAR does not require agencies to seek full and open competition for contracts awarded using the simplified acquisition procedures (used for contracts that do not exceed the simplified acquisition threshold—generally $150,000 or $7 million when acquiring commercial items), though agencies are required to compete them to the maximum extent practicable. The National Defense Authorization Act for Fiscal Year 2017 raised the simplified acquisition threshold to $250,000. Pub. L. No. 115-91, § 805. In accordance with special emergency procurement authorities, DHS and DOD increased the simplified acquisition threshold for the three 2017 hurricanes to $750,000. DHS and DOD exercised the use of the special emergency procurement authorities within 41 U.S.C. §1903, as amended by §816 of the National Defense Authorization Act for Fiscal Year 2017, Pub. L. No. 114-328 (and as implemented by DHS FAR Class Deviation 17-02 and DOD Class Deviation 2017-O0007, respectively) to increase the simplified acquisition threshold for procurements in support of the three 2017 hurricanes to $750,000.

When it becomes evident that a base contract period and any option periods will expire before a subsequent contract to meet the same need can be awarded, contracting officers may, for example,

- extend the existing contract, or
- award a short-term stand-alone contract to the incumbent contractor on a non-competitive basis to avoid a lapse in services, along with sufficient justification and approval.

These extensions and new sole source contracts are informally referred to as bridge contracts by some in the acquisition community, and we use that terminology in this chapter. In October 2015, we established the following definitions related to bridge contracts:

- Bridge contract: An extension to an existing contract beyond the period of performance (including base and option years), and a new, short-term contract awarded on a sole-source basis to an incumbent contractor to avoid a lapse in service caused by a delay in awarding a follow-on contract.[21]
- Predecessor contract: The contract that was in place prior to the award of a bridge contract.
- Follow-on contract: A longer-term contract that follows a bridge contract for the same or similar services. This contract can be competitively awarded or awarded on a sole-source basis.

Contracts, orders, and extensions (both competitive and non-competitive) are included in our definition of a "bridge contract" because the focus of the definition is on the intent of the contract, order, or extension.[22]

---

[21] GAO, *Sole Source Contracting: Defining and Tracking Bridge Contracts Would Help Agencies Manage Their Use*, GAO-16-15 (Washington, D.C.: Oct. 14, 2015).

[22] GAO's definition includes all types of extensions, both those extensions of contracts that may be considered "competitive," e.g., the use of FAR 52.217-8 when it was evaluated at award, and those that are "noncompetitive," e.g., those extensions of requirements that are used to

However, the FAR does not formally define bridge contracts or require that they be tracked. We recommended that the Office of Federal Procurement Policy amend the FAR to incorporate a definition of bridge contracts. The Office of Federal Procurement Policy agreed with our recommendation to provide guidance to agencies on bridge contracts and has taken steps to develop that guidance, but has not yet implemented our recommendations.[23]

If a contracting officer opts to extend the existing contract in place—often referred to as a predecessor contract—the contracting officer may use a number of different mechanisms to do this. One of these is the "option to extend services" clause. If the contract includes this clause, the contracting officer may use it to extend the contract for up to six months.[24] While this option may be exercised more than once, the total extension of performance shall not exceed 6 months.

---

extend the period of performance beyond that of the original contract (and all the options) and require a justification and approval, when the intention is to bridge a gap in services.

[23] The Office of Federal Procurement Policy has not yet taken action on our recommendations. In the National Defense Authorization Act for Fiscal Year 2018 (Pub. L. No. 115-91), Congress established a definition of bridge contracts that is applicable to DOD requirements owners, meaning members of the armed forces (excluding Coast Guard) or a civilian employee of the Department of Defense responsible for a requirement for a service to be performed through a services contract. Specifically, Section 851 defined a bridge contract as (1) an extension to an existing contract beyond the period of performance to avoid a lapse in service caused by a delay in awarding a subsequent contract; or (2) a new short-term contract awarded on a sole-source basis to avoid a lapse in service caused by a delay in awarding a subsequent contract. Section 851 requires that by October 1, 2018, the Secretary of Defense is to ensure that a requirements owner must to the extent practicable, plan appropriately before the date a service is needed to avoid the use of a bridge contract for services. DOD issued a bridge contracts memorandum in January 2018, which defined bridge contracts as modifications to existing contracts to extend the period of performance, increase the contract ceiling or value or both, or a new interim sole-source contract awarded to the same or a new contractor to cover the timeframe between the end of the existing contract and the award of a follow-on contract.

[24] FAR § 17.208(f) provides for the use of the clause cited at FAR § 52.217-8, "Option to Extend Services" in solicitations and contracts for services when the inclusion of an option is appropriate.

FEMA AND USACE RELIED ON ADVANCE CONTRACTS TO RESPOND TO THE 2017 DISASTERS, BUT FEMA LACKS AN UPDATED ADVANCE CONTRACTING STRATEGY AND GUIDANCE

FEMA and USACE obligations on advance contracts—as of May 31, 2018—accounted for about half of total federal contract obligations for the three hurricanes, and more than three quarters of the contract obligations identified by those agencies for the California wildfires. However, an outdated strategy and lack of guidance to contracting officers resulted in confusion about whether and how to prioritize and use advance contracts to quickly mobilize resources in response to the three 2017 hurricanes and the California wildfires.

Advance Contracts Accounted for about Half of Government-wide Contract Obligations for the 2017 Hurricanes, and over Three-Quarters of FEMA and USACE's Obligations for the California Wildfires

Government-wide contract obligations for the three hurricanes were about $8.2 billion as of May 31, 2018.[25] FEMA and USACE obligated 46 percent, or about $3.8 billion, of the $8.2 billion spent government-wide on the three hurricanes through advance contracts. Data on government-wide

---

[25] Federal agencies may continue to contract for goods and services well after a disaster has occurred. For example, federal agencies are still making contract obligations as part of recovery efforts as far back as Hurricanes Sandy and Katrina, which occurred in 2012 and 2005, respectively. In addition to DOD and DHS, the following departments had contract obligations in support of hurricanes Harvey, Irma, and Maria response efforts: the departments of Agriculture, Commerce, Energy, Health and Human Services, Housing and Urban Development, Interior, Justice, Labor, State, Transportation, Treasury, and Veterans Affairs; the Agency for International Development; the Broadcasting Board of Governors; the Corporation for National and Community Service; the Environmental Protection Agency; the Federal Communications Commission; the General Services Administration; the National Aeronautics and Space Administration; the National Archives and Records Administration; the Small Business Administration; and the Social Security Administration.

contract obligations for the California wildfires were not able to be identified because national interest action codes were not established for them in FPDS-NG.[26] However, FEMA and USACE provided information on their contracting activities related to the wildfires. Their use of advance contracts accounted for 86 percent, or about $667 million, of the contract obligations they identified.

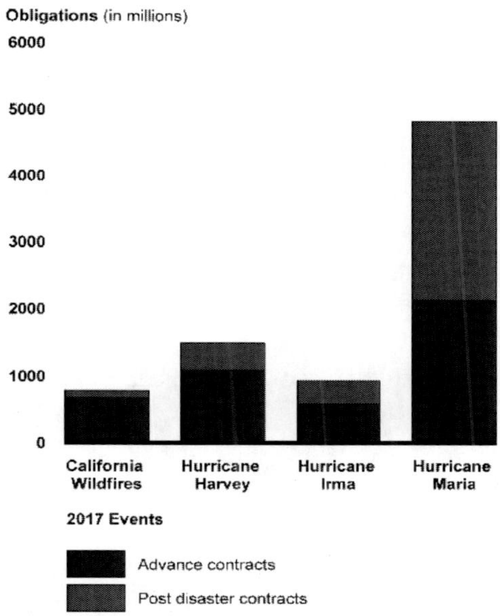

Source: GAO analysis of Federal Procurement Data System-Next Generation data. | GAO-19-93.

Notes: Obligations are nominal 2017 and 2018 dollars.

FEMA officials told us that contracts awarded in support of the U.S. Virgin Islands after Hurricane Maria impacted the islands may all be coded under Hurricane Maria, regardless of whether the contract responds to needs from Hurricane Irma or Hurricane Maria.

Figure 2. Federal Emergency Management Agency (FEMA) and U.S. Army Corps of Engineers Contract Obligations in Response to the Three 2017 Hurricanes and California Wildfires, as of May 31, 2018.

---

[26] National interest action codes are used to track certain contracts in support for disaster response and recovery efforts in FPDS-NG. No national interest action code was established for the 2017 California wildfires.

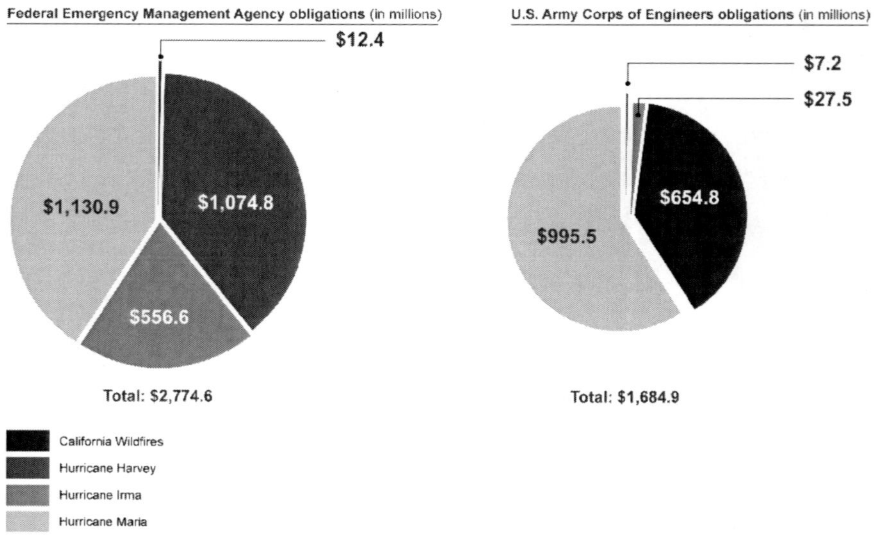

Source: GAO analysis of Federal Procurement Data System-Next Generation data. | GAO-19-93.

Note: Obligations are nominal 2017 and 2018 dollars. Numbers may not add due to rounding.

Figure 3. Federal Emergency Management Agency and U.S. Army Corps of Engineers Advance Contract Obligations in Response to the Three Hurricanes and California Wildfires in 2017, as of May 31, 2018.

FEMA and USACE advance contract obligations for the three hurricanes and California wildfires totaled about $4.5 billion, about 56 percent of the total contract obligations made by these agencies for these disasters. See Figure 2 for details on FEMA and USACE's advance and post-disaster contract obligations by event. The greatest proportion of FEMA and USACE's obligations on advance contracts supported Hurricane Maria disaster relief efforts—41 percent and 59 percent, respectively. About 39 percent of USACE's obligations on advance contracts were used in support of the California wildfires, compared to less than 1 percent of FEMA's obligations. FEMA awarded orders against 72 base advance contracts in response to the three 2017 hurricanes and California wildfires, and USACE awarded orders against 15 of its advance contracts. See Figure 3 for FEMA and USACE's obligations on advance contracts by event.

Advance Contracts Were Used Primarily for Services

FEMA and USACE procured a variety of goods and services through advance contracts in response to the three hurricanes and wildfires, but about 86 percent of obligations, or $3.8 billion, were used to procure services. For example, all of USACE's $1.7 billion in advance contract obligations were for services, such as debris removal. FEMA obligated about $2.2 billion on services, such as architect and engineering services to rebuild roads and bridges.

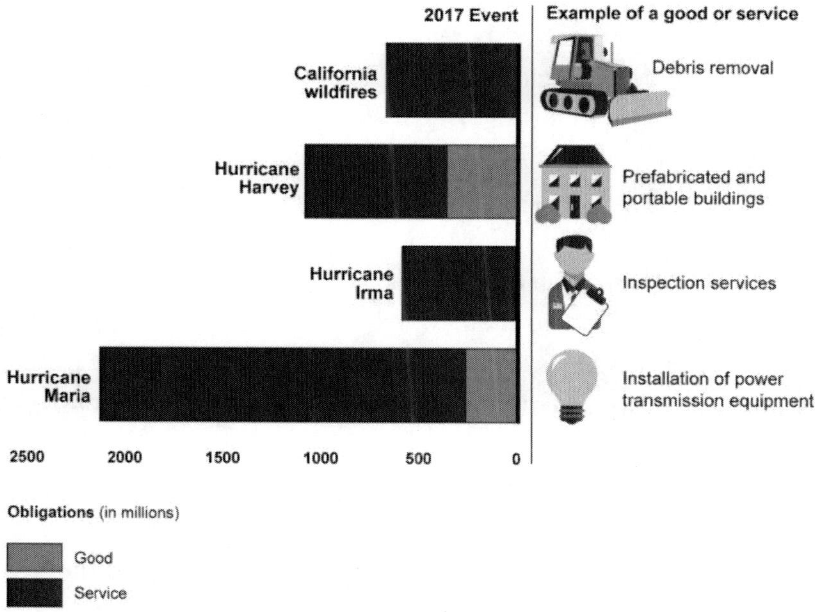

Source: GAO analysis of Federal Procurement Data System-Next Generation data. | GAO-19-93.

Notes: Obligations are nominal 2017 and 2018 dollars.

FEMA officials told us that contracts awarded in support of the U.S. Virgin Islands after Hurricane Maria impacted the islands may all be coded under Hurricane Maria, regardless of whether the contract responds to needs from Hurricane Irma or Hurricane Maria.

Figure 4. Federal Emergency Management Agency (FEMA) and U.S. Army Corps of Engineers Advance Contract Obligations and Examples of Goods and Services Used to Respond to the 2017 Hurricanes and Wildfires, as of May 31, 2018.

FEMA's obligations on goods totaled $624 million and included prefabricated buildings, such as manufactured housing units to provide lodging, and food and water. See Figure 4 for examples of obligations on goods or services by event.

FEMA Lacks an Updated Strategy and Guidance on the Use of Advance Contracts

FEMA lacks an updated strategy and guidance on advance contract use, despite the PKEMRA requirement to develop a contracting strategy that maximizes their use to the extent practical and cost effective. As we found in May 2006 following Hurricane Katrina, and reiterated in our September 2015 report, agencies need to have competitively awarded contracts in place before a disaster to be effective in their response. Our current review found that FEMA has established advance contracts for goods and services to enable it to respond following a disaster. However, FEMA's lack of an updated strategy and guidance on advance contract use resulted in confusion about whether and how to maximize their use to the extent cost-effective and practical to facilitate a faster response when providing goods and services to survivors.

PKEMRA required the FEMA Administrator to identify specific goods and services that the agency could contract for in advance of a natural disaster in a cost-effective manner.[27] PKEMRA also required the FEMA Administrator to develop a contracting strategy that maximizes the use of advance contracts to the extent practical and cost-effective. Following the enactment of PKEMRA, in 2007 FEMA issued the *Advance Contracting of Goods and Services Report to Congress*, in part to address the requirement for an advance contracting strategy.[28] In addition to the strategy, FEMA provides information on advance contracts in its *Disaster Contracting Officer Desk Guide*.

---

[27] Pub. L. No. 109-295, § 691 (codified at 6 U.S.C. § 791).

[28] Department of Homeland Security, Federal Emergency Management Agency, *Advance Contracting of Goods and Services Report to Congress*.

The 2007 strategy notes that advance contracts will help to preclude the need to procure goods and services for disaster response under the unusual and compelling urgency exception to full and open competition, and allow FEMA to rapidly mobilize resources in immediate response to disasters. Several contracting officials we spoke with said that it is a requirement to use advance contracts before awarding new contracts. Moreover, a senior FEMA contracting official told us that advance contracts are intended to be used before awarding post-disaster contracts, even if the advance contract is not capable of fulfilling all of the requirements for a needed good or service. However, our review of the strategy found that it does not provide any specific direction on how contracting officers should award or use advance contracts to meet PKEMRA's objectives, or how they should be prioritized in relation to post-disaster contracts.[29] Further, there is no mention in FEMA's 2017 *Disaster Contracting Officer Desk Guide* that advance contracts should be considered prior to the award of post-disaster contracts.

In September 2015, we found shortfalls with the information available to contracting officers about advance contracts and recommended that FEMA provide new or updated guidance with information on how advance contracts should be used.[30] FEMA agreed with this recommendation and stated that in 2015 it included information on advance contracts and their use in training documentation. However, our review of semi-annual training documentation provided in May 2018 found that it only lists some of the advance contracts that are available, and not guidance on their use.

A report by the Senate Committee on Homeland Security and Governmental Affairs identified concerns about FEMA's use of advance contracts for self-help tarps in response to the 2017 hurricanes. Specifically, the report found that while FEMA ordered some tarps through one of its existing advance contracts, that order was placed after a post-

---

[29] The 2007 strategy states that FEMA developed an acquisition tracker to prioritize contracts across four tiers, based on their internal and external visibility and risk. However, the strategy states the tracker only identifies contracts used in 2007 and 2008, and FEMA contracting officials we spoke with were not familiar with the acquisition tracker, and said it had been replaced by FEMA's advance contract list, which has no prioritization.

[30] GAO-15-783.

disaster contract for tarps was signed, raising questions about whether FEMA's actions were informed by an overall strategy for using its advance contracts, in this case, for tarps. Our current review identified similar concerns, and found that the lack of an updated strategy and guidance on the use of advance contracts contributed to challenges in using these contracts to respond to the 2017 disasters.

In our review of advance contracts for meals and tarps, we found the following:

- **Meals:** Prior to the 2017 disasters, FEMA had advance contracts in place to provide meals with specific nutritional requirements. According to FEMA contracting officials, the advance contract vendors were at capacity for these specific meals following the response to Hurricane Harvey, requiring FEMA to issue a new post-disaster competitive solicitation and award new contracts with less specific nutritional requirements following Hurricane Maria. Based on our review of contract documentation, two of the existing advance contract vendors were awarded these new post-disaster contracts, but at different prices than those negotiated through their advance contracts. FEMA officials told us that contracting officers will negotiate to ensure the price of the contract is fair and reasonable and may utilize historical information or current contract prices to inform this determination. Normally, adequate price competition establishes a fair and reasonable price. According to a contracting officer involved with the award, FEMA relied on competition and historical prices, but not the existing advance contract prices, to determine that the new post-disaster contract meal prices were fair and reasonable. Guidance on the extent to which advance contract prices should be considered when comparing proposed prices to historical prices paid could help to further inform contracting officers' decision-making during a disaster.
- **Tarps:** Our review of FEMA's use of contracts for tarps is another example of how FEMA lacked an updated advance contracting

strategy and guidance to provide goods and services to facilitate a faster response to the 2017 disasters. For example, in September 2014, FEMA awarded multiple award indefinite delivery, indefinite quantity advance contracts to three small businesses for self-help tarps, which are used to cover small areas of roof damage. In November 2014, these contracts were modified by the contracting officer to include delivery requirements for providing tarps to replenish FEMA's stock during steady state operations or during emergency response operations, such as a natural disaster. The contract modification added that during an emergency response, vendors would be expected to deliver up to 150,000 tarps within 96 hours of being issued a task order. However, these small businesses were not required to meet the emergency response delivery time frames and amounts since they would not be expected to store tarps on FEMA's behalf, limiting the use of FEMA's advance tarp contracts for immediate disaster response needs. According to a contracting officer involved with these contracts, the tarp advance contracts are typically used only to replenish tarp stockpiles in FEMA's distribution centers. However, the contracting officer also noted that not being able to fully use the existing advance contracts for tarps to respond to the three 2017 hurricanes was a challenge and required FEMA to award post-disaster contracts to meet tarp requirements.

Furthermore, we found that FEMA awarded post-disaster contracts for tarps before utilizing its advance contracts with the small businesses. Contract file documentation for the post-disaster contracts stated that FEMA's advance contract holders for tarps had reached their capacity, and that market research had confirmed that it would be difficult for small businesses to meet the urgent delivery timeframes for tarps. Yet, after the award of the post-disaster tarp contracts, FEMA awarded task orders to one of the advance contractors to provide tarps in response to Hurricane Maria. Another small business advance contractor, which according to FEMA's post-disaster contract documentation had reached its capacity, also submitted a proposal as part of the post-disaster contract solicitation. According to FEMA, neither of the post-disaster contract holders ultimately provided the required tarps.[31] The timing and use of the existing tarp advance contracts raises questions about their ability to provide tarps immediately following a disaster, and whether an updated advance contracting strategy would have enabled FEMA to more quickly provide the needed tarps to survivors, considering the additional time and staff resources needed to award new post-disaster contracts.

FEMA established advance contracts to provide critical goods, like meals and tarps, following a disaster; however FEMA's 2007 contracting strategy does not provide direction on the objectives of advance contracts or how to maximize their use to the extent practical and cost-effective, as required by PKEMRA. According to FEMA officials, they had not considered updating the 2007 advance contracting strategy because they believed the use of advance contracts following PKEMRA had been incorporated into their disaster contracting practices. FEMA has also not communicated specific guidance to program and contracting officials on whether and how advance contracts should be prioritized before issuing new post-disaster solicitations and awarding contracts for the same or similar requirements, or how to maximize their use to the extent practical and cost-effective following a disaster, as required by PKEMRA. FEMA officials also acknowledged that additional guidance regarding advance contracts, including their availability and use during a disaster, could be useful. Without an updated strategy—and clear guidance that is incorporated into training—on the use of advance contracts and how they should be prioritized and used in relation to new post-disaster contract awards, FEMA lacks reasonable assurance that it is maximizing the use of advance contracts to quickly and cost-effectively provide goods and

[31] The first post-disaster contract was terminated for convenience citing shortages in available tarps following the 2017 hurricane season. A stop work order was issued for the second post-disaster contract following concerns over whether the tarps received met FEMA's specifications.

services following a disaster. This places FEMA at risk of continued challenges in quickly responding to subsequent disasters.

IMPROVEMENTS NEEDED IN FEMA'S PLANNING, MANAGEMENT, AND REPORTING OF ADVANCE CONTRACTS

While FEMA used a variety of advance contracts to respond to the 2017 disasters, we found weaknesses in the process of awarding and overseeing selected advance contracts in our review. These weaknesses were: (1) challenges in FEMA's acquisition planning; (2) limited record keeping or management of certain FEMA contracts; and (3) incomplete reporting on FEMA's advance contract actions to certain congressional committees. Related to USACE, we did not identify any planning or management challenges based on our review of its four selected contracts, and USACE is not required to report on its advance contract actions to the congressional committees.

Challenges in FEMA's Acquisition Planning Resulted in Bridge Contracts

FEMA has taken some steps since 2016 to improve competition and develop processes and guidance on the acquisition process for advance contracts, but shortfalls in acquisition planning have resulted in a number of bridge contracts. Bridge contracts can be a useful tool in certain circumstances to avoid a gap in providing products and services. We have previously reported that when non-competitive bridge contracts are used frequently or for prolonged periods, the government is at risk of paying more than it should for products and services.[32]

---

[32] GAO-16-15.

Based on our analysis, 63 of FEMA's 72 advance contracts used in response to the 2017 disasters were initially competed. All 15 of USACE's advance contracts used in responding to the three hurricanes and California wildfires in 2017 were initially competed. We found that at least 10 of FEMA's advance contracts used in 2017 were bridge contracts.[33] Within the 10 FEMA advance contracts we identified as bridge contracts, 6 were part of our selected case studies. The six advance contracts with subsequent bridges in our review obligated roughly $778 million in response to the three hurricanes and California wildfires in 2017.[34] These bridge contracts included five that are associated with two of FEMA's largest programs used in 2017—the Individual Assistance Program and Public Assistance Program—and one that is associated with a telecommunications program.

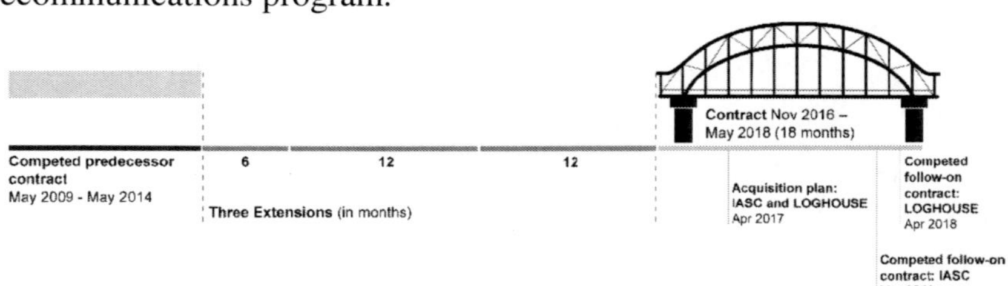

LOGHOUSE – Logistics Housing Operations Unit Installation, Maintenance, and Deactivation Contracts

IASC – Individual Assistance Support Contract

Source: GAO graphic based on information from contract file reviews and interviews with contracting officials. | GAO-19-93.

Note: The Individual Assistance Support Contract (IASC) provides the mass care and temporary housing portion of the Individual Assistance program. The Logistics Housing Operations Unit Installation, Maintenance, and Deactivation (LOGHOUSE), provides manufactured housing units to support the Logistics Manufactured Housing Unit program.

Figure 5. Timeline of the Individual Assistance-Technical Assistance Contract III Bridge.

---

[33] As previously noted, bridge contracts occur when an agency realizes that it will not be able to award a competed or non-competed follow-on contract before the current contract expires and can be done either by extending the current contract's period of performance or by awarding a new short term contract to the incumbent vendor(s) to avoid a lapse in service.

[34] These six contracts all had additional bridges related to the same requirement, though not all bridge contracts we identified were non-competitive.

Three of the six bridge advance contracts we reviewed were awarded to support FEMA's Individual Assistance Program, which provides mass care services such as food and water as well as financial and direct assistance, among other services, to survivors whose property has been damaged or destroyed and whose losses are not covered by insurance. In 2017, this assistance was supported through the Individual Assistance-Technical Assistance Contract (IA-TAC), known as IA-TAC III. The IA-TAC III predecessor contracts had an original period of performance from a base year starting in May 2009 with four 1-year options that ended in May 2014. However, FEMA program and contracting officials were unable to implement changes to the requirements—recommended by FEMA senior leadership in 2010—prior to expiration. According to FEMA officials, staffing shortfalls, operational tempo, and unrealistic contract requirements led to acquisition planning delays. These challenges, in turn, led to a series of extensions from May 2014 to November 2016 and a new non-competitive bridge contract (base with options) from November 2016 to May 2018. At that point new, competitive follow-on indefinite delivery indefinite quantity contracts—the Individual Assistance Support Contract (IASC) and Logistics Housing Operations Unit Installation, Maintenance, and Deactivation (LOGHOUSE)—were awarded. See Figure 5.

Two of our six selected advance contracts that were bridge contracts were awarded to support FEMA's Public Assistance Program, which provides supplemental federal assistance to state, tribal, territorial, and local governments for debris removal, life-saving emergency protective measures, and the repair, replacement, or restoration of damaged facilities. The predecessor Public Assistance-Technical Assistance Contract (PA-TAC) used in 2017, known as PA-TAC III, was awarded with an original period of performance from a base year in February 2012 with four 1-year options that ended in February 2017. FEMA officials noted that changes to the PA-TAC III contract requirements and acquisition strategy were identified in 2015. Yet due to the time needed to incorporate these changes, FEMA was unable to complete required acquisition planning activities, such as finalizing the acquisition plan, prior to the expiration of PA-TAC III. Following 11 months of extensions to complete these activities, FEMA competitively awarded new contracts in December 2017. These awards were protested to the GAO and the protests were denied and are currently under review at the Court of Federal Claims. According to FEMA officials, these events required PATAC III to be extended until January 2019, as shown in Figure 6.

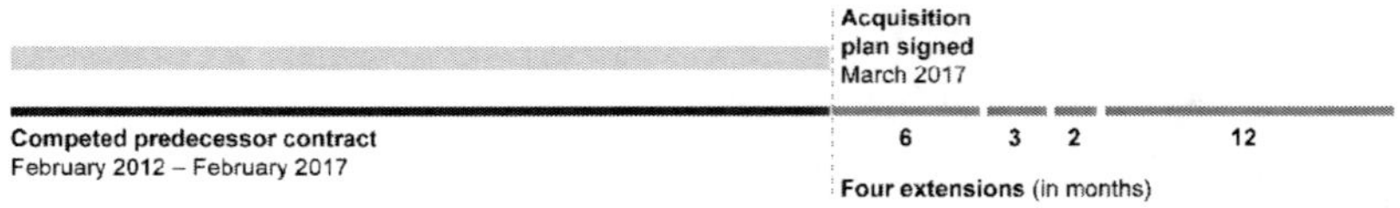

Source: GAO graphic based on information from contract file reviews and interviews with contracting officials. | GAO-19-93.

Figure 6. Timeline of Public Assistance-Technical Assistance Contract III Bridge.

The remaining bridge contract in our sample is associated with the Wireline Services Program, a telecommunication program that provides FEMA employees deployed to respond to a disaster with local and long-distance telephone, high-speed data, and cable television services. The 5 year wireline predecessor contract was awarded in 2003 and again in 2008, but FEMA was unable to award a competed contract when the 2008 contract expired in December 2013 due to the time it took to update program requirements. FEMA contracting officials extended the contract for 6 months before letting it expire altogether. Due to high staff turnover and inconsistent record keeping, at the time of our review FEMA officials were unable to determine the cause for this lapse of service, which occurred after the contract's expiration in June 2014. Starting in January 2015, FEMA contracting officials used a series of bridge contracts over more than three years to address changing contract requirements and delays in completing acquisition planning documentation, as shown in Figure 7. FEMA contracting officials anticipated awarding a competitive contract by the end of fiscal year 2018, but the award has been delayed and the existing contract extended through January 2019.

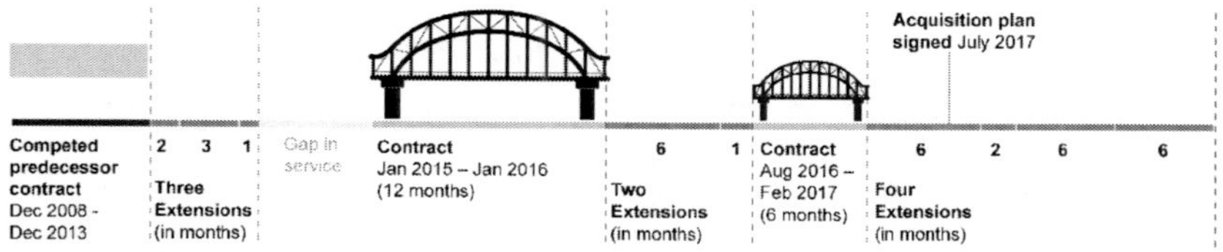

Source: GAO graphic based on information from contract file reviews and interviews with contracting officials. | GAO-19-93.

Figure 7. Timeline of Wireline Bridge.

In one of the bridge contracts included in our review, FEMA improperly used FAR clause 52.217-8. According to that clause, an agency may extend a contract's period of performance for up to 6 months and is generally used in the event of circumstances outside of the contracting officer's control that prevent the new contract award, such as a bid protest. This clause may be used multiple times to extend the contract so long as the total extension of performance does not exceed 6 months. Our analysis found that FEMA used the clause for a total of 14 months to justify two 6-month extensions and one 2-month extension to the second bridge contract. The FEMA contracting official associated with the advance contract reported uncertainty over the proper use of this clause and what other authorities should have been used instead to extend the contract. FEMA's Office of Chief Counsel and contracting officials acknowledged this error.

While not all bridge contracts that we identified during our review were non-competitive, FEMA officials acknowledged that the use of non-competitive bridge contracts is not an ideal practice as they cannot ensure the government is paying what it should for products and services. In October 2015 we identified delays in the completion of acquisition planning documentation as one of the leading causes of awarding bridge contracts.[35] In an effort to decrease the need for non-competitive bridge contracts and provide ample time for acquisition planning, FEMA began implementing a 5-Year Master Acquisition Planning Schedule (MAPS) in 2016. MAPS is a tracking tool that monitors the status of and provides

---

[35] GAO-16-15. Acquisition planning documentation can include an independent government cost estimate, the statement of work, acquisition plan, market research, and evaluation criteria depending on contract value.

acquisition planning timeframes for certain FEMA acquisitions over $5 million, as well as for all advance contracts and any acquisition deemed by the agency to be mission critical, regardless of dollar value.

As we previously noted, acquisition planning includes both the pre-solicitation and solicitation phases. Based on our review of MAPS documentation, the tool generates a timeline of discretionary acquisition milestones across these two phases, based on certain considerations like the type of acquisition and whether it will be competed. Using this timeline, MAPS sends email alerts to program and contracting staff when certain acquisition milestones should occur.

Specific to the solicitation phase, FEMA's Office of the Chief Procurement Officer has developed annual lead time guidance for how long contracting officers should be given to award new contracts following the completion of the acquisition package, which is then conveyed through MAPS. For example, for acquisitions $150,000 and under, FEMA's 2018 lead time guidance states contracting officers should be given 60 days to award the contract following completion of the acquisition package. FEMA officials we spoke with acknowledged that these discretionary timeframes are frequently shortened when program office officials are delayed in completing acquisition packages.

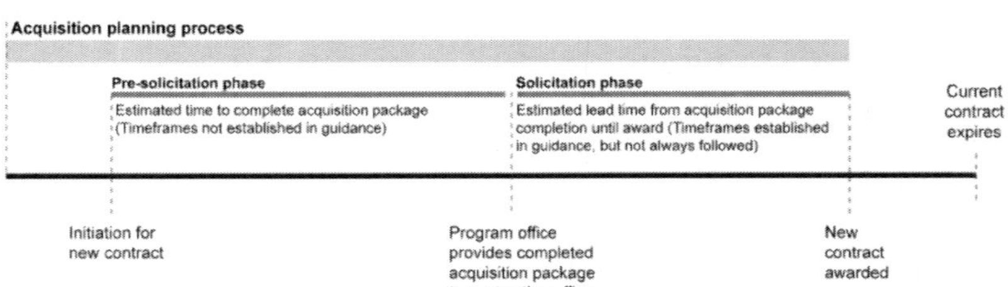

Source: GAO depiction of milestones identified in FEMA's 5 Year Master Acquisition Planning Schedule. | GAO-19-93.

Figure 8. Notional Depiction of Acquisition Planning Timeline Depicted in FEMA's 5 Year Master Acquisition Planning Schedule (MAPS).

While FEMA has lead time guidance to establish timeframes for completing the solicitation phase, FEMA currently has no guidance establishing timeframes for the pre-solicitation phase, when program offices complete the acquisition packages. Figure 8 provides an example timeline of the major milestones tracked in MAPS. In its analysis of 12 fiscal year 2017 contracts tracked in MAPS that were awarded late, FEMA found that half were late because contracting officials were not given enough lead time to award a new contract following the program office's completion of the acquisition package. Not adhering to suggested timeframes can place a burden on contracting officers and increase the likelihood of not awarding the contract on schedule, requiring FEMA to non-competitively extend the existing contract.

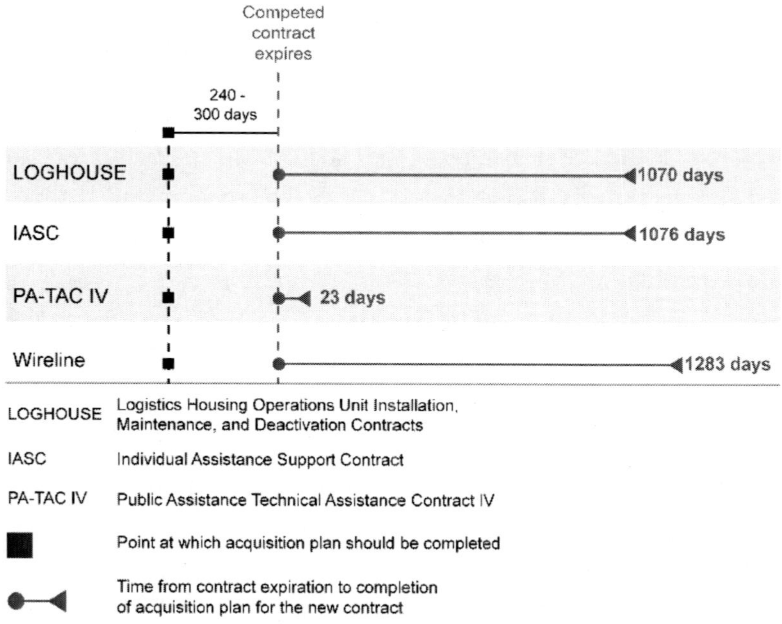

Source: GAO analysis of signed acquisition plans for identified bridge contracts. | GAO-19-93.

Note: The acquisition package can consist of documents such as an independent government cost estimate, statement of work, acquisition plan, and market research among others. The applicable documents are determined by dollar value.

Figure 9. Acquisition Plan Completion Compared to Competed Contract Expiration.

According to FEMA's lead time guidance, based on the contract values for the bridge contracts in our review contracting officers should have been given between 240 and 300 days to award a new contract once the acquisition package was completed. However, as we mention earlier, due to delays from changing program requirements and acquisition strategies we found that the acquisition plans for the follow-on contracts related to these bridge contracts were not completed until after the predecessor contract had already expired, as shown in Figure 9 below. Timely completion of the acquisition package was a key challenge identified in the contracts we reviewed. However, according to officials from the Office of the Chief Procurement Officer, they do not have the authority to establish guidance for FEMA program officials on completing pre-solicitation phase activities. In August 2011, we identified challenges with acquisition planning across DHS. Specifically, we found that DHS and other agencies did not measure or incorporate into guidance the amount of time it takes to develop and obtain approvals of the acquisition planning documents required during the pre-solicitation phase.[36] We recommended that DHS procurement offices collect information about the timeframes needed for the acquisition planning process to establish timeframes for when program officials should begin acquisition planning. DHS did not concur with this recommendation, stating that its acquisition manual already encourages early planning, and has not implemented the recommendation. At the time, we maintained that program officials needed more guidance to have a better understanding of how much time to allow for completing acquisition planning steps, and that the component procurement offices are best positioned to provide guidance on how long these planning processes may take. Given the current challenges we identified with FEMA's ability to complete acquisition planning activities in a timely manner and the resulting delays in awarding new contracts for critical advance contract goods and services, additional information and guidance on acquisition planning timeframes remains important.

Additionally, while MAPS has been in place since 2016 and FEMA officials have instituted training to communicate the system's intent, program and contracting officials we spoke with varied in their familiarity

---

[36] GAO, *Acquisition Planning: Opportunities to Build Strong Foundations for Better Services Contracts*, GAO-11-672 (Washington D.C.: Aug. 9, 2011).

with it. For example, officials responsible for MAPS stated that by March 2016, 90 percent of FEMA's contracting staff had attended an hour long training session and additional training sessions were held for all program office staff at various points in 2016 and 2017. However, most of the program office and contracting officials responsible for the bridge contracts in our review reported limited familiarity with MAPS.

While FEMA has taken some positive steps to institute training and has guidance on timeframes for part of the acquisition planning process, program and contracting staff we spoke with were still uncertain how best to utilize MAPS to identify the time needed to effectively complete acquisition planning activities. According to federal internal control standards, agency management should internally communicate the necessary quality information to achieve their objectives.[37] Given FEMA's emphasis on planning before a disaster and using advance contracts to help reduce the need to award non-competitive contracts during a disaster, establishing clear guidance on the factors that can affect acquisition planning activities, and requiring officials to follow the timeframes needed to complete them to meet the goal of awarding competitive contracts, is essential. Until FEMA provides detailed guidance about timeframes and considerations that affect the entire acquisition planning process—both the pre-solicitation and solicitation phases—to all officials responsible for acquisition planning, and clearly communicates the intent of MAPS, it cannot ensure that MAPS will be effective at reducing the number of non-competitively awarded bridge contracts, as is FEMA's intent.

Current Record-Keeping Practices Limit Visibility into Advance Contract Management

While FEMA has procedures regarding the documentation required for its contract files, current practices limited visibility into the advance contracts in our review. Specifically we found that acquisition plans and

[37] GAO, *Standards for Internal Control in the Federal Government*, GAO-14-704G (Washington, D.C.: September 2014).

some other contract documents were unable to be located in certain cases. Acquisition plans provide the program and contract history as well as other information on which acquisition decisions, such as the type of contract required, are based.

FEMA contracting officials were unable to locate acquisition plans for 4 of our 10 FEMA selected advance contracts despite FAR and DHS acquisition guidance requiring plans for these particular contracts to be completed and stored in the contract file.[38] Three of these acquisition plans are associated with the IA-TAC bridge contract which, as previously noted, was associated with one of FEMA's largest programs used in 2017. FEMA contracting officials were also unable to locate the acquisition plans completed for the prior iteration of IA-TAC because they were not in the hard copy contract file or contract writing system, meaning that no acquisition plan guiding the IA-TACs since before its 2009 award could be found. In 2011, the DHS Office of the Inspector General conducted a review of FEMA's IA-TAC and identified, among other things, incomplete contract files as a problem.[39] Not being able to locate acquisition plans can result in the loss of contract knowledge and lessons learned from prior awards.

Additionally, we found instances of contract documentation for advance contracts related to our case studies that contract officials could not locate. For instance, FEMA was unable to confirm whether or not an option year for the last competed Wireline contract included in the contract was exercised due to a lack of documentation. In order to obtain this answer, FEMA officials had to reach out to the vendor for their records. Moreover, the modification exercising the first option year for one of the IA-TAC III predecessor contracts was missing, as were the determination and findings documents exercising the first option year for all three of the predecessor IA-TAC III contracts that were associated with the advance

[38] FAR § 7.103(e) and Department of Homeland Security, *Department of Homeland Security Acquisition Manual* § 3007.103(e) (Oct. 2009).

[39] Department of Homeland Security Office of the Inspector General, *Improving FEMA's Individual Assistance, Technical Assistance Contracts* (Washington D.C.: Sept. 27, 2011).This chapter found, among others, that pricing memorandums—which help contracting officers determine if a vendor's proposed price is fair and reasonable—for two of the IA-TACs were missing from the contract files and unable to be located.

contracts in our review. After we made FEMA officials aware of the missing documentation, they subsequently added clarifying memos to the contract files.

FEMA standard operating procedures state that the acquisition documents in the official contract file will be sufficient to constitute a complete history of the entire transaction for the purpose of providing a complete background, and as a basis for informed decisions at each step in the acquisition process. Additionally, these procedures require headquarters staff to place modifications to contracts and orders and associated supporting documentation in the contract file within 5 business days of awarding a contract or issuing an order. FEMA officials stated they are required to follow these procedures until DHS has fully transitioned to an electronic filing system. According to DHS officials, that system is currently in the testing phase and a timeframe for implementation has not yet been finalized. Furthermore, according to these officials, DHS has not yet decided which, if any, existing contracts will be required to be retroactively entered into the system. Until this decision has been made and implementation occurs, FEMA's official file of record for its advance contracts consists of a hardcopy file, which contracting officers at FEMA headquarters are required to add completed contract documentation to, per the standard operating procedures.[40] A FEMA official told us that some documentation, including some of the missing documentation we identified, has been lost due to staff turnover and an office move in 2016.

FEMA officials anticipate some of the challenges associated with managing the hard copy advance contract files will be alleviated after implementation of the Electronic Contract File System. However, DHS officials have not decided whether components will be required to retroactively enter contract information for any contract awarded prior to the implementation date. This would require FEMA and other DHS components to continue to maintain hardcopy files for some contracts—including large strategic sourcing vehicles and advance contracts—for the

[40] While the official file of record remains hardcopy, FEMA officials said they highly encourage the use of its current contract writing system as an unofficial means of electronically storing necessary documentation. Although this is highly encouraged, FEMA officials noted they cannot require contracting officials to use this system for electronic contract file storage.

foreseeable future. For example, FEMA's $2.7 billion LOGHOUSE, and $14 million IASC advance contracts were awarded in 2018 and have a period of performance lasting until 2023. Until FEMA adheres to existing contract file management requirements, whether the contract files will be transferred into the electronic system or remain in hard copy format, it is at continued risk of having incomplete contract files and a loss of institutional knowledge regarding these advance contracts.

Information on Advance Contracts in FEMA's Disaster Contract Quarterly Reports to Congressional Committees Is Incomplete

Since December 2007, FEMA has submitted quarterly reports to congressional committees that list all disaster contracting actions in the preceding three months.[41] These quarterly reports also include details on contracts awarded by non-competitive means, as required by PKEMRA.[42] However, our analysis shows that some reports from fiscal year 2017 and 2018 have been incomplete. In September 2015, we found that FEMA's quarterly reports to congressional committees in fiscal years 2013 and 2014 did not capture all of FEMA's noncompetitive orders.[43] At that time, FEMA attributed this to an error in data compilation prior to mid-2013 and explained that it had updated its process for collecting these data and strengthened the review process, resulting in accurate reports starting in the fourth quarter of fiscal year 2013. Despite this change in the data collection process, our current analysis found that 29 contract actions associated with the 10 selected advance contracts in our review were not reported across

[41] DHS officials stated that in addition to its quarterly reports there are other methods, such as the congressional notification tool, that provide information on contract actions to the appropriations committees. However, these other methods are not in direct response to the PKEMRA requirement that FEMA report on all contracts awarded by non-competitive means.

[42] For the purpose of its congressional committee reports, FEMA considers a contract action to be an action that uses appropriated funds designated for disaster situations, excluding any that deobligate funds or constitute no-cost modifications.

[43] GAO-15-783.

FEMA's fourth quarter fiscal year 2017 and first quarter fiscal year 2018 reports. For example, FEMA's fourth quarter fiscal year 2017 report did not include 13 contract actions equaling about $83 million, or 15 percent, of the $558 million in total obligations associated with the 10 selected advance contracts in our review. Similarly, FEMA's first quarter fiscal year 2018 report did not include 16 contract actions equaling about $122 million, or 23 percent, of the $532 million in total obligations associated with the 10 selected advance contracts in our review. Figure 10 provides a breakdown of the total contract action obligations by extent of competition.

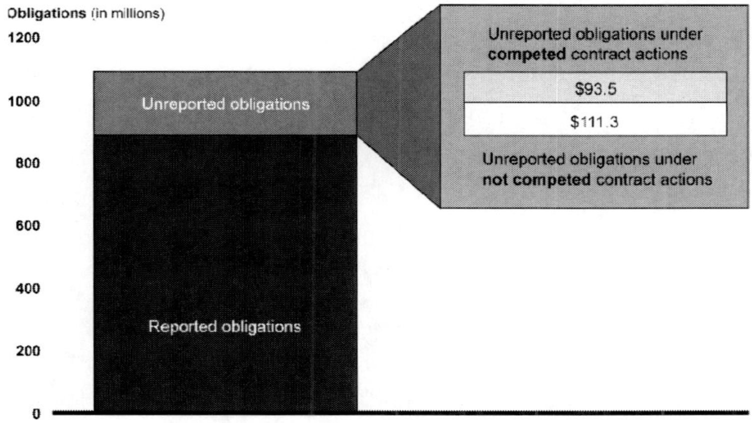

Source: GAO analysis of FEMA Fourth Quarter Fiscal Year 2017 and First Quarter Fiscal Year 2018 Quarterly Reports to Congress. | GAO-19-93.

Note: For the purpose of its congressional committee reports, FEMA considers a contract action to be an action that uses appropriated funds designated for disaster situations, excluding any that deobligate funds or constitute no-cost modifications.

Figure 10. Total Value of Contract Action Obligations from Selected Advance Contracts in Fourth Quarter Fiscal Year 2017 and First Quarter Fiscal Year 2018 Federal Emergency Management Agency (FEMA) Congressional Committee Reports.

To compile the quarterly reports, FEMA officials told us that their methodology is to pull contract action data that is documented in their contract writing system and FPDS-NG roughly one week after the end of each fiscal quarter. Once the data are pulled from these two sources, officials said they compare the data to ensure all reported actions are captured. However, according to officials, the data may not include all contract actions. Specifically, during disaster response efforts like those in 2017, FEMA policy allows contracting officers to execute what it refers to as "notice to proceed," which is a notice to a construction contractor to begin work under certain circumstances. FEMA officials responsible for the quarterly reports stated that if notice to proceed documentation is used, information on some contract actions that were issued during the fiscal quarter, but not entered into the systems until after the quarter ended, may be missed during the data compilation process.

FEMA policy requires that contracting officers who execute the notice to proceed documentation complete the contract award documentation in the contract writing system within three days of when the contracting officer receivers the contractor's acceptance of the notice. However, a FEMA policy official acknowledged that during disaster response, this does not always occur. Further, FEMA officials responsible for compiling the reports stated that it is not part of their methodology to review data from prior fiscal quarters to see whether any contract actions have been entered that were not previously reported. By not adhering to FEMA policy that establishes timeframes for entering data in a disaster response scenario, FEMA risks reporting incomplete information. Moreover, without taking steps to ensure its reporting methodology provides complete information on all competed and not competed disaster contract actions, FEMA cannot be certain it is providing the congressional committees with visibility into all of its overall disaster contract awards or the extent of non-competitive contract obligations over time.

No Challenges Identified with the Planning and Management of Selected USACE Advance Contracts

The four selected USACE advance contracts in our review—one supporting USACE's temporary power mission and three supporting its debris removal mission—were awarded in 2014 with a period of performance lasting until 2019. Since these contracts have not reached the end of their period of performance, we were unable to assess the effectiveness of USACE planning activities. According to contracting officials, USACE is performing acquisition planning activities for both the temporary power and debris removal advance contracts and anticipates awarding the new contracts prior to the current contracts' expiration.

Additionally, USACE was able to provide the acquisition plans for each of the four advance contracts in our review. Unlike FEMA, which retains hard copy files of its contract documentation, USACE uses three official systems of record to store contract file documentation electronically. Officials acknowledged that while moving between the three official systems to find documents may be time consuming, contract documents are typically able to be located.

FEMA and USACE Identified Lessons Learned from the Use of Advance Contracts in 2017, but Reported Challenges with State and Local Coordination Remain

Both FEMA and USACE have processes for identifying and assessing lessons learned following a disaster. Contracting officials from these agencies identified several lessons learned from the 2017 major hurricanes and the California wildfires that directly affected their use of advance contracts. These include the need for: (1) additional advance contracts for certain goods and services; (2) flexibility to increase contract ceilings; (3) use of USACE's debris removal advance contracts to respond to the California wildfires; and (4) federal coordination and information sharing with state and local governments on advance contracts. While officials identified some lessons learned, they also identified challenges related to FEMA's outreach with state and local governments on advance contracting efforts.

FEMA and USACE Have Identified Lessons Learned and Actions to Address Them

FEMA and USACE have processes for identifying and assessing lessons learned through after-action reviews and reports following major disasters. According to FEMA and USACE officials, they routinely perform these reviews and then compile after-action reports to identify lessons learned and proposed actions to address them. Due to the concurrent nature of hurricanes Harvey, Irma, and Maria, FEMA headquarters completed one combined after-action review for all three hurricanes in July 2018.[44] The resulting report identified 18 strategic-level key findings across five focus areas, and recommendations for improvement.[45] These recommendations included some that were specific to advance contracts, such as the need for additional advance contracts to support future disaster response efforts, and improved state and local coordination to support state and local contracting and logistics operations. In addition, USACE officials performed after-action reviews following disasters, and have a process in place to discuss challenges and recommendations for improvement on their use of advance contracts for temporary power, temporary roofing, and debris removal.

While the scope of FEMA's and USACE's after-action reports are broader than just advance contracts, we identified, based on our review of reports and interviews with FEMA and USACE officials, several lessons learned related to advance contracts following the 2017 hurricanes and California wildfires, as shown in Table 1.

---

[44] FEMA, *2017 Hurricane Season FEMA After-Action Report* (July 12, 2018).

[45] The five focus areas are: scaling a response for concurrent, complex incidents; staffing for concurrent, complex incidents; sustained whole community logistics operations; responding during long-term infrastructure outages; mass care to initial housing operations.

Table 1. Lessons Learned and Proposed Actions Regarding Advance Contracts Identified by the Federal Emergency Management Agency (FEMA) and the U.S. Army Corps of Engineers (USACE) Following the 2017 Hurricanes and California Wildfires

Issue	Lessons Learned	Proposed Actions
Additional advance contract requirements	Both FEMA and USACE reported a need for additional advance contracts in certain areas following their response to the 2017 disasters. For example, FEMA officials explained that following the disaster response efforts for Hurricane Maria, they realized there was a need for additional advance transportation contracts, specifically for freight and barge transportation, to carry goods to areas outside the continental United States, like Puerto Rico and the U.S. Virgin Islands. USACE also reported a shortage of advance contracts for quality assurance services—inspection of the services by contractors— which officials said required them to spend extra time and resources awarding post-disaster contracts for these services.	FEMA officials confirmed that they awarded an additional advance contract for freight and barge transportation in August 2018, but stated that the contract award was being protested. USACE officials told us they are considering whether to pursue additional advance contracts for quality assurance services.
Contract ceilings	FEMA and USACE officials noted that as a result of the magnitude and timeframes of the 2017 disasters, ceilings on a number of indefinite delivery, indefinite quantity advance contracts were too low to support response and recovery efforts and were increased through modification. For example, FEMA had modified its bottled water and telecommunications advance contracts to increase the dollar value ceilings. USACE officials reported increasing ceilings for their debris removal and temporary power contracts.	FEMA and USACE officials said they have taken steps to update contract ceiling language to provide greater flexibility. For example, in August 2018, FEMA issued an alert to contracting officers to insert language into solicitations and contracts when issuing multiple award indefinite delivery, indefinite quantity contracts that allows the contracting officer to incrementally increase the contract's dollar value ceiling during a presidentially declared disaster so long as the total amount of supplies and services ordered does not exceed the total capacity of all vendors under the contract able to provide the needed goods and services. USACE officials said they have revised language

Table 1. (Continued)

Issue	Lessons Learned	Proposed Actions
		in solicitations for their indefinite delivery, indefinite quantity temporary power and debris removal contracts to include higher estimated cost ceilings than those in prior advance contracts.
Debris removal pricing	USACE officials identified challenges using their debris removal advance contracts to respond to the 2017 California wildfires. According to USACE officials, the debris removal contract requirements are written broadly enough to include all kinds of debris removal, but they have primarily been used to manage hurricane debris removal. As a result, the advance contracts did not address some of the issues related to the 2017 wildfire debris removal. For example, according to USACE officials, while hurricane debris is often priced by volume, officials stated that the wildfire debris had to be priced by tonnage, and wrapped to prevent contamination.	As of August 2018, USACE officials stated that they are in the process of developing ordering instructions at the task order level based on the prices and experiences from the 2017 wildfires. Specifically, officials stated that because the unique circumstances of a disaster vary, they are developing ordering instructions that will reflect historical prices from using their debris removal contracts to respond to the wildfires that can be leveraged at the task order level in the newly awarded debris removal contracts in April 2019.

Source: GAO summary of FEMA and USACE information. I GAO-19-93.

Challenges in Coordinating with and Providing Information to State and Local Governments on the Use of Advance Contracts Continued

We also found that while FEMA has updated its guidance to reflect some requirements for state and local coordination over the use of advance contracts, inconsistencies in FEMA's outreach and information on the use of advance contracts remains a challenge. PKEMRA required that FEMA encourage state and local governments to establish their own advance contracts with vendors for goods and services in advance of natural disasters.[46] In September 2015, we found that FEMA's outreach with state and local governments to encourage the establishment of advance contracts can result in more efficient contracting after a disaster.[47] PKEMRA also required that FEMA establish a process to ensure that federal advance contracts are coordinated with state and local governments, as appropriate. In our September 2015 report, we also found that these efforts can ensure that states are aware of and can access certain federal advance contracts, such as General Services Administration schedule contracts.

However, in the same report, we found that inconsistencies in whether and how the regions perform state and local outreach limited FEMA's ability to support advance contracting efforts.[48] We recommended that FEMA provide new or updated guidance to ensure that all contracting officers are aware of requirements concerning the need to conduct outreach to state and local governments to support their use of advance contracts. DHS concurred with this recommendation and in 2017 FEMA updated its *Disaster Contracting Desk Guide* to state that contracting officers should inform their state and local counterparts of the availability and use of federal advance contracts established by FEMA.[49] Our review of the guide found that it does remind contracting officers to coordinate with states and localities over the use of federal advance contracts, but does not provide any details on how often or what types of advance contract information should be shared with states and localities, or provide any instructions to contracting officers on PKEMRA's requirement to encourage states and localities to establish their own advance contracts for the types of goods and services needed during a disaster.

Our current review also found inconsistencies with FEMA's efforts to encourage states and localities to establish their own advance contracts with vendors and ensure coordination with them on their use of federal advance contracts. For example, some regional FEMA officials explained that they regularly perform outreach, which can assist states and localities with establishing advance contracts for goods and services commonly

---

[46] Pub. L. No.109-295, § 691 (codified at 6 U.S.C. § 791).
[47] GAO-15-783.
[48] GAO-15-783.
[49] FEMA Office of the Chief Procurement Officer, *Disaster Contracting Desk Guide 2017*.

needed during a disaster, like security, transportation, and office supplies. Regional officials we spoke with said more frequent coordination allows them to avoid overlap across state and federal contracting efforts, and know what resources the states have in place and how long states are capable of providing these resources following a disaster. However, other regional officials reported having less frequent coordination with state and local governments. For example, a FEMA official stated that one of the regions has less frequent meetings with state and local governments because the region is geographically dispersed and has fewer disasters. According to another regional official, coordination between some regional offices and state and local officials over advance contracting was minimal prior to Hurricane Harvey, and in some cases only occurred when FEMA and state and local officials were co-located during a disaster.

Officials from some state and local governments and USACE reported examples where increased coordination between FEMA, states, and localities could have improved the use of advance contracts in 2017. For example, in September 2018 we found that some localities were relying on the same contractors to perform debris removal activities following Hurricanes Harvey in Texas and Irma in Florida.[50] As a result, we reported that some contractors that were removing debris in Texas did not honor existing contracts in Florida, leading to delays in debris removal. Additional communication and coordination between FEMA and contracting officials in these states and localities about which contractors they had established advance contracts with could have helped to prevent this overlap and subsequent delay in removing debris. During our current review, USACE and California officials also reported miscommunications about state and local expectations for USACE's debris removal contracts following the wildfires. Specifically, USACE and state and local officials reported differing expectations about the work to be performed under USACE's debris removal contracts, such as what structures would be removed from private property and acceptable soil contamination levels. According to USACE officials, they relied on FEMA, as the lead for

---

[50] GAO-18-472, *2017 Hurricanes and Wildfires: Initial Observations on the Federal Response and Key Recovery Challenges*, (Washington, D.C.: Sept. 4, 2018).

coordinating federal disaster response, to manage communication with states and localities and to identify and manage expectations about the scope of work to be performed using their advance debris removal contracts. While state and local officials we met with in California reported working closely with some FEMA officials not responsible for regional contracting during the response to the wildfires, FEMA regional contracting officials said that they had no direct coordination with California officials.

We also identified inconsistencies in the information available to FEMA's contracting officials on existing advance contracts, which can be used to facilitate coordination with states and localities on the establishment and use of advance contracts. Our review of FEMA's advance contract list found that it does not include all of the advance contracts that FEMA has in place, and contracting officers we spoke with cited other resources they also use to identify advance contracts, like biannual training documentation provided to contracting staff. For example, while FEMA officials told us the advance contract list is updated on a monthly basis, our analysis found that 58 advance contracts identified on the June 2018 advance contract list were not included in the May 2018 biannual training documentation, including contracts for telecommunications services, generators, and manufactured housing units. Further, 26 of the contracts included in the May training documentation were not included on the June advance contract list, including contracts for foreign language interpretation services, hygiene items, and short-shelf life meals. Some contracting officers we spoke with said they referred to the advance contract list as the primary resource for identifying advance contracts, while others referenced the biannual training as their primary resource.

FEMA has recognized some shortcomings in how it coordinated and communicated with state and local governments over the use of advance contracts following the 2017 disasters, and identified some action to address these issues moving forward. In the *2017 Hurricane Season FEMA After-Action Report*, FEMA identified the need to expand its capabilities to support state, local, tribal, and territorial governments in improving their capabilities for advance contracting, among other issues.[51] The report recommends that FEMA should continue efforts to develop a toolkit that will provide state and local governments with recommendations for advance contracts, emergency acquisition guidance, and solicitation templates.

According to FEMA contracting officials, the development of the toolkit has been prioritized by FEMA's Administrator to help better prepare the states and localities and decrease their reliance on FEMA for assistance following a disaster. However, as of August 2018 the specific contents of the toolkit were still being decided. For example, officials familiar with the development of the toolkit originally said they intended for it to include FEMA's advance contract list, to provide states with recommendations on the types of advance contracts that may be useful. But in subsequent discussions these officials told us they did not plan to provide states and localities with a full list of advance contracts to avoid being overly prescriptive, and because not all of the contracts on the list are relevant for the types of disasters some states experience. Officials further stated that since it is the responsibility of the federal coordinator in each region to communicate available federal advance contracts to states and localities, providing a full list of advance contracts is unnecessary.

Federal internal control standards state that agency management should use quality information to achieve their objectives.[52] Agency management should also internally and externally communicate that information to achieve their objective. However, FEMA's guidance does not clearly communicate its objectives and requirements for contracting officers to encourage states and localities to enter into their own advance contracts, nor is there a consolidated resource listing available advance contracts that states and localities can use to inform their advance contracting efforts. According to FEMA officials, information on advance contracts is fluid, as new contracts are established or old contracts expire. Officials also told us that the advance contract list is updated monthly, yet as mentioned earlier, contracts identified in the May training

---

[51] FEMA, *2017 Hurricane Season FEMA After-Action Report* (July 12, 2018).
[52] GAO-14-704G.

documentation were not reflected in the list that was updated as of June. Ensuring that advance contract information is complete and updated regularly is important, because differences across FEMA's resources listing advance contracts could result in FEMA's contracting officers not being aware of the availability of certain contracts during a disaster, and states not receiving recommendations on what advance contracts may be helpful for them to establish. Without clear guidance on FEMA's expectations for coordination with states and localities on advance contracting efforts, and a centralized resource listing up to date information on FEMA's advance contracts, FEMA contracting officers and their state and local counterparts lack reasonable assurance they will have the tools needed to effectively communicate about advance contracts, and use them to respond to future disasters. Moreover, given FEMA's recent emphasis on the importance of states and localities having the capability to provide their own life-saving goods and services in the immediate aftermath of a disaster, clearly communicating consistent and up to date information on the availability and limitations of federal advance contracts through the toolkit, or other means, is critical to informing state and local disaster response efforts.

CONCLUSION

Contracting during a disaster can pose a unique set of challenges as officials face a significant amount of pressure to provide life-sustaining goods and services to survivors as quickly as possible. Advance contracts are a tool that FEMA and others within the federal government can leverage to rapidly and cost-effectively mobilize resources, while also helping to preclude the need to procure critical goods and services non-competitively after a disaster. Given the circumstances surrounding the 2017 disasters and the importance of preparedness for future disasters, it is critical to ensure that the federal government is positioned to maximize its advance contracts to the extent practical and cost-effective to provide immediate disaster response.

Although FEMA has identified advance contracts for use during a disaster, without an updated strategy—and guidance that is incorporated into training—on how to maximize their use during a disaster, as well as the development of clear guidance on acquisition planning timeframes, FEMA is at risk of these contracts not being effectively planned and used. Furthermore, FEMA officials have not always maintained complete information on the advance contracts available for them to quickly respond to disasters, or completely reported competitively and non-competitively awarded advance contract information to better help congressional committees evaluate spending over time. Finally, without continued efforts to improve outreach with states and localities and centralize information on available advance contracts, FEMA's contracting officers and their state and local counterparts may not have the information needed to efficiently respond to a disaster.

RECOMMENDATIONS FOR EXECUTIVE ACTION

We are making nine recommendations to FEMA.

FEMA's Administrator should update the strategy identified in its 2007 *Advance Contracting of Goods and Services Report to Congress* to clearly define the objectives of advance contracts, how they contribute to FEMA's disaster response operations, and whether and how they should be prioritized in relation to new post-disaster contract awards. (Recommendation 1)

FEMA's Administrator should ensure the Head of the Contracting Activity updates the *Disaster Contracting Desk Guide* to include guidance for whether and under what circumstances contracting officers should consider using existing advance contracts prior to making new post-disaster contract awards, and include this guidance in existing semiannual training given to contracting officers. (Recommendation 2)

FEMA's Administrator should update and implement existing guidance for program office and contracting officer personnel to identify acquisition planning timeframes and considerations across the entire acquisition planning process, and clearly communicate the purpose and use of MAPS. (Recommendation 3)

FEMA's Administrator should ensure the Head of the Contracting Activity adheres to current hard copy contract file management requirements to ensure advance contract files are complete and up to date, whether they will be transferred into the new Electronic Contract Filing System or remain in hard copy format. (Recommendation 4)

FEMA's Administrator should ensure the Head of the Contracting Activity reminds contracting officers of the three day timeframe for entering completed award documentation into the contract writing system when executing notice to proceed documentation. (Recommendation 5)

FEMA's Administrator should ensure the Head of the Contracting Activity revises its reporting methodology to ensure that all disaster contracts are included in its quarterly reports to congressional committees on disaster contract actions. (Recommendation 6)

FEMA's Administrator should ensure the Head of the Contracting Activity revises the *Disaster Contracting Officer Desk* guide to provide specific guidance for contracting officers to perform outreach to state and local governments on the use and establishment of advance contracts. (Recommendation 7)

FEMA's Administrator should ensure the Head of the Contracting Activity identifies a single centralized resource listing its advance contracts and ensure that source is updated regularly to include all available advance contracts. (Recommendation 8)

FEMA's Administrator should ensure the Head of the Contracting Activity communicates information on available advance contracts through the centralized resource to states and localities to inform their advance contracting efforts. (Recommendation 9)

AGENCY COMMENTS AND OUR EVALUATION

We provided a draft of this chapter to DOD, DHS, and FEMA for review and comment. DOD did not provide any comments on the draft report. In its comments, reprinted in appendix IV, DHS and FEMA concurred with our nine recommendations. DHS and FEMA also provided technical comments, which we incorporated as appropriate.

In its written comments, FEMA agreed to take actions to address our recommendations, such as updating guidance on advance contract use and management, adding an addendum to its quarterly report that captures the contract actions that were previously unreported, and better communicating information on advance contracts to states and localities. In its concurrence with two of our recommendations, FEMA requested that we consider these recommendations resolved and close as implemented based on our actions it had previously taken.

For example, in its response to our third recommendation, FEMA agreed to update and implement existing guidance to identify acquisition timeframes and the purpose and use of its 5-Year MAPS program. In its response, FEMA reiterated that it has conducted training sessions for its contracting and program staff on the 5-Year MAPS program and provides notice to program managers when acquisition planning is set to begin, which the agency believes satisfies this recommendation. We acknowledge FEMA's training in this chapter; however, we noted that not all program and contracting staff we spoke with were familiar with 5-Year MAPS, and there is no formal guidance on timeframes for the entire acquisition planning process. We continue to believe this recommendation remains open and encourage FEMA to formalize guidance on the timeframes and considerations for planning various types of acquisitions across the entire acquisition planning process, and to document the purpose and use of the 5-Year MAPS program to ensure a uniform understanding of the program.

Further, in its concurrence with our eighth recommendation, FEMA stated that it believes its current advance contract list satisfies our recommendation for internally communicating available advance contracts. We acknowledge in this chapter that the advance contract list is updated monthly; however, we found inconsistencies in the advance contract list and other documentation identifying advance contracts, which could result in FEMA's contracting officers not having full visibility into available advance contracts. We continue to believe the recommendation remains open and encourage FEMA to identify a centralized resource with all available advance contracts and ensure that it is regularly updated for contracting staff.

We are sending copies of this chapter to the appropriate congressional committees, the Secretary of Defense, the U.S. Army Corps of Engineers Director of Contracting, the Secretary of Homeland Security, the Administrator of the Federal Emergency Management Agency, and the Federal Emergency Management Agency's Chief Procurement Officer.

Marie A. Mak
Director, Contracting and National Security Acquisitions

List of Requesters

The Honorable Michael B. Enzi
Chairman
Committee on the Budget
United States Senate

The Honorable Ron Johnson
Chairman

The Honorable Claire McCaskill
Ranking Member
Committee on Homeland Security and Governmental Affairs
United States Senate

The Honorable Rand Paul, MD
Chairman

The Honorable Gary Peters
Ranking Member
Subcommittee on Federal Spending Oversight and Emergency Management
Committee on Homeland Security and Governmental Affairs
United States Senate

The Honorable Jeb Hensarling
Chairman

The Honorable Maxine Waters
Ranking Member
Committee on Financial Services
House of Representatives

The Honorable Michael T. McCaul
Chairman

The Honorable Bennie G. Thompson
Ranking Member
Committee on Homeland Security
House of Representatives

The Honorable Trey Gowdy
Chairman

The Honorable Elijah E. Cummings
Ranking Member
Committee on Oversight and Government Reform
House of Representatives

The Honorable Bill Shuster
Chairman

The Honorable Peter DeFazio
Ranking Member
Committee on Transportation and Infrastructure
House of Representatives

The Honorable Daniel M. Donovan, Jr.
Chairman
Subcommittee on Emergency Preparedness, Response, and Communications
Committee on Homeland Security
House of Representatives

The Honorable Sean Duffy
Chairman

The Honorable Emanuel Cleaver, II
Ranking Member
Subcommittee on Housing and Insurance
Committee on Financial Services
House of Representatives

The Honorable Gary J. Palmer
Chairman
Subcommittee on Intergovernmental Affairs
Committee on Oversight and Government Reform
House of Representatives

The Honorable Ann Wagner
Chairman

The Honorable Al Green
Ranking Member
Subcommittee on Oversight and Investigations
Committee on Financial Services
House of Representatives

The Honorable Nydia M. Velazquez
House of Representatives

APPENDIX I: OBJECTIVES, SCOPE, AND METHODOLOGY

This chapter reviews the federal government's contracting efforts for preparedness, response, and recovery efforts related to the three 2017 hurricanes and California wildfires. This chapter specifically addresses the use of advance contracts, assessing the extent to which (1) the Federal Emergency Management Agency (FEMA) and the U.S. Army Corps of Engineers (USACE) used advance contracts, (2) the planning, management, and reporting of selected FEMA and USACE advance contracts met certain contracting requirements, and (3) FEMA and USACE identified any lessons learned and challenges with their use of these contracts. We also have an ongoing review on post-disaster contracting that is expected to be completed in early 2019. To identify the extent to which FEMA and USACE used advance contracts, we reviewed data on contract obligations for the 2017 disasters from the Federal Procurement Data System-Next Generation (FPDS-NG) through May 31, 2018.[53] We identified hurricane obligations using the national interest code, as well as the contract description. Data on obligations for the California wildfires is limited to those contracts that FEMA and USACE identified as being used to respond to those events because no national interest code was established in FPDS-NG.[54] To determine which obligations were made through the use of advance contracts, we reviewed documentation provided by FEMA and USACE identifying the advance contracts they have in

---

[53] For the purposes of this chapter, contract obligations include obligations against what the General Services Administration's FPDS-NG categorizes as definitive vehicles (definitive contracts and purchase orders that have a defined scope of work that do not allow for individual orders under them), and against what FPDS-NG categorizes as indefinite delivery vehicles (orders under the Federal Supply Schedule, orders/calls under blanket purchase agreements, orders under basic ordering agreements, orders under government-wide acquisition contracts, and orders under other indefinite delivery vehicles, such as indefinite delivery, indefinite quantity contracts).

[54] National interest action codes are used to track relief contracts in FPDS-NG. No national interest action code was requested for the 2017 California wildfires.

place and that were used in support of the 2017 disasters.[55] We analyzed the FPDS-NG data to identify FEMA and USACE advance contract obligations compared to overall contract obligations by disaster, competition procedures used, and the types of goods and services procured. We assessed the reliability of FPDS-NG data by reviewing existing information about the FPDS-NG system and the data it collects—specifically, the data dictionary and data validation rules—and performing electronic testing. We determined the FPDS-NG data were sufficiently reliable for the purposes of this chapter.

To assess the extent to which FEMA used its advance contracts, we reviewed FEMA contracting policies and guidance, such as FEMA's 2017 *Disaster Contracting Desk Guide* and FEMA's *Advance Contracting of Goods and Services Report to Congress* to identify available guidance on the use and intent of advance contracts. Based on our review of documentation, we identified examples of goods—tarps and meals—that FEMA had advance contracts in place for, but experienced challenges using in response the 2017 disasters. We reviewed FPDS-NG data to determine whether these goods were procured through post-disaster contracts rather than advance contracts, and selected advance and post-disaster contracts for further review. To identify limitations that affected the use of tarp and meal advance contracts, we gathered and reviewed advance and post-disaster contract documentation and interviewed contracting officials involved in the award and use of the contracts in 2017.

To assess the extent to which the planning, management, and reporting of advance contracts used in response to the three hurricanes and California wildfires in 2017 met selected applicable contracting requirements, we reviewed relevant documentation, including the Post-Katrina Emergency Management Reform Act (PKEMRA), the Federal Acquisition Regulation (FAR), and Department of Homeland Security

[55] DHS and DOD exercised the use of the special emergency procurement authorities within 41 U.S.C. §1903, as amended by section 816 of the National Defense Authorization Act for Fiscal Year 2017, Pub. L. No. 114-328 (and as implemented by DHS FAR Class Deviation 17-02 and DOD Class Deviation 2017-O0007, respectively) to increase the micro-purchase threshold to $20,000 for procurements in support of these major disaster responses so contract obligations for hurricanes reported in FPDS-NG may only include obligations over that amount.

(DHS, FEMA, and USACE contracting policies. We identified a non-generalizable sample of advance contracts based on advance contract obligation data from FPDS-NG as of March 31, 2018. We analyzed the data to identify 10 competed and four h non-competed contracts.[56] To obtain a range of competed contracts, we identified contracts used for goods and services with obligations above $50 million. All of the non-competed contracts used were for FEMA services; to obtain a range of non-competed contracts we identified contracts with obligations above $10 million. Our selected advance contracts included 10 from FEMA and four from USACE. Findings based on information collected from the 14 contracts cannot be generalized to all advance contracts. Additional details on our selected contracts can be found in Table 2.

To review our selected FEMA and USACE advance contracts, we developed a data collection instrument to gather selected contract information, such as period of performance, contract type, estimated contract value, and the presence of key contract documents, among others. To assess FEMA and USACE's planning of selected advance contracts, we reviewed information from our data collection instrument on advance contract award date and period of performance, and determined that six of FEMA's contracts met GAO's definition of a bridge contract. To identify any planning challenges that contributed to these extensions, we reviewed FEMA acquisition planning policies, timeframes and relevant contract file documentation, such as written acquisition strategies and justification and approval documents, to determine whether acquisition planning activities for the selected advance contracts were completed according to guidance. We interviewed FEMA officials associated with these contracts on acquisition planning efforts, and factors that affected their ability to award new contracts. We also reviewed documentation and interviewed officials on FEMA's acquisition planning system—the 5 Year Master Acquisition Planning Schedule (MAPS).

[56] For the purposes of selecting our case studies, competitive contracts included contracts and orders coded in FPDS-NG as "full and open competition," "full and open after exclusion of sources," and "competed under simplified acquisition procedures." Noncompetitive contracts included contracts and orders coded in FPDS-NG as "not competed," "not available for competition," and "not competed under simplified acquisition procedures."

Table 2. Selected Federal Emergency Management Agency (FEMA) and U.S. Army Corps of Engineers (USACE) Advance Contracts

	Contract Number	Description	Obligations as of May 31, 2018
FEMA	HSFE7017D0001 HSFE7017D0003 HSFE7017D0004	Non-competed service contract supporting FEMA's individual assistance program which provides financial assistance to those with property damaged by a disaster.	$100,525,716 $39,082,653 $249,035,558
	HSFE3016D0203	Non-competed service contract which provides FEMA employees with local and long-distance telephone service, high speed data services, and cable television that can be used during disaster response.	$13,963,151
	HSFE7014D0152 HSFE7014D0150	Competed goods contract for manufactured housing units to shelter those displaced by a disaster.	$74,332,500 $70,294,193
	HSFEHQ12D0881 HSFEHQ12D0882	Competed service contract supporting FEMA's public assistance program which provides supplemental federal assistance to state, tribal, territorial, and local governments for disaster response.	$197,464,836 $177,798,052
	HSFE7017D0019	Competed goods contract for bottled water to distribute to those affected by a disaster.	$154,627,988
	HSFE8013D0070	Competed services contract which provides housing inspection services for those with property damaged by a disaster.	$321,518,435
USACE	W911WN15D0001	Competed services contract which provides temporary power for households affected by a disaster.	$718,436,308
	W912P814D0016 W912P814D0028 W912P814D0033	Competed services contract which provides debris removal for households affected by a disaster.	$166,540,852 $321,687,964 $320,351,681

Source: GAO review of Federal Procurement System Database- Next Generation and contract information I GAO-19-93.

To assess FEMA and USACE's management of selected advance contracts, we reviewed information gathered from our data collection instrument on the presence of selected acquisition documents, such as acquisition strategies and contract modifications in the contract file, that typically provide the history of a contract. We reviewed relevant procurement regulations, the DHS Acquisition Manual, and other FEMA and USACE policies, to identify acquisition documentation requirements and record keeping processes. For contracts where documentation was not found in the contract file or system of record, we requested the missing documentation from FEMA and USACE officials to determine whether it had been completed. We also interviewed FEMA and USACE headquarters officials to supplement our understanding of FEMA and USACE's record keeping policies, practices, and challenges.

To assess the reporting of selected advance contracts, we compared advance contract action data identified in FPDS-NG to data reported in *FEMA's Disaster Contracts Quarterly Report Fourth Quarter, Fiscal Year 2017* and *Disaster Contracts Quarterly Report First Quarter, Fiscal Year 2018* to congressional committees on disaster contracting to identify any unreported actions. We interviewed FEMA officials to discuss the methodology and data sources for the congressional committee reports, and any limitations to the accuracy of the data reported.

To assess what challenges and lessons learned FEMA and USACE identified with the use of advance contracts in 2017, we reviewed PKEMRA advance contract requirements, FEMA and USACE documentation on the use of advance contracts, and after-action reports from 2017 and prior years, including the *Hurricane Sandy FEMA After-Action Report,* and *the 2017 Hurricane Season FEMA After-Action Report*, and federal internal control standards for information and communications. As part of our review, we identified FEMA and USACE's processes for documenting lessons learned following a disaster, lessons learned specific to advance contracts, and any recommendations or actions planned by the agencies to address them. We interviewed FEMA and USACE headquarters officials on reported lessons learned, any other challenges related to the use of advance contracts, and ongoing or completed actions to address them. To describe challenges related to coordination with state and local officials on the use of advance contracts, we interviewed FEMA and USACE regional staff. To obtain perspectives and examples from state and local government officials involved in disaster response efforts we interviewed officials in California on advance contracting efforts. The information gathered from these officials is not generalizable to all officials. We also analyzed information on available advance contracts from FEMA's June 2018 advance contract list and FEMA's May 2018 training documentation identifying advance contracts to identify any differences in the information available to FEMA regional contracting officers, and their state and local contracting counterparts.

We conducted this performance audit from March 2018 to December 2018 in accordance with generally accepted government auditing standards. Those standards require that we plan and perform the audit to obtain sufficient, appropriate evidence to provide a reasonable basis for our findings and conclusions based on our audit objectives. We believe that the evidence obtained provides a reasonable basis for our findings and conclusions based on our audit objectives.

APPENDIX II: FEDERAL EMERGENCY MANAGEMENT AGENCY (FEMA) REGIONAL OFFICES

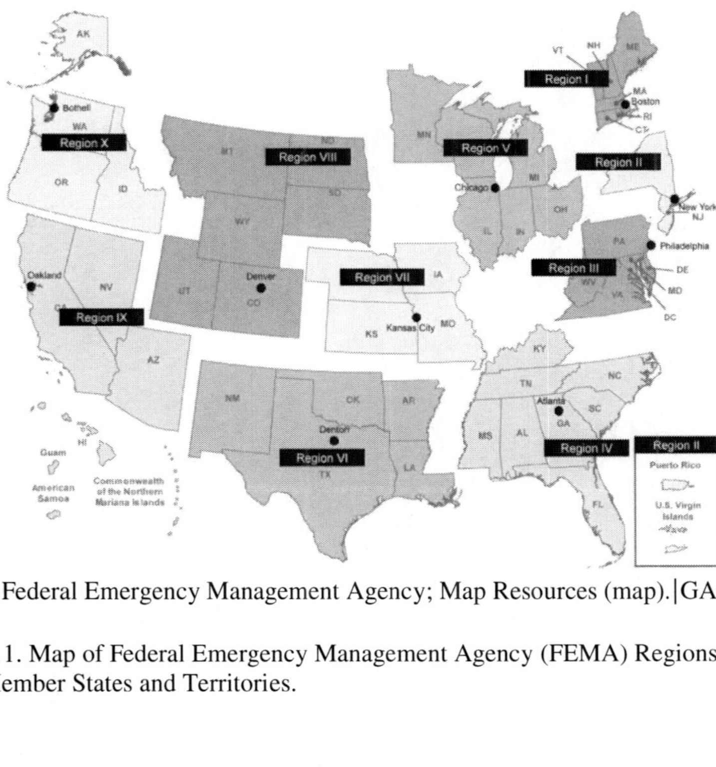

Source: Federal Emergency Management Agency; Map Resources (map).|GAO-19-93.

Figure 11. Map of Federal Emergency Management Agency (FEMA) Regions and Their Member States and Territories.

APPENDIX III: FEDERAL EMERGENCY MANAGEMENT AGENCY (FEMA) AND U.S. ARMY CORPS OF ENGINEERS (USACE)-IDENTIFIED ADVANCE CONTRACTS

Table 3. Federal Emergency Management Agency (FEMA) Identified Advance Contracts, as of June 2018

Acquisition Name	Description	Number of Contracts	Procurement Vehicle	Base Award Date
Short-term Lodging Assistance Program	Allows FEMA to use Corporate Lodging Consultant services to place disaster victims in hotels during mass evacuations.	1	GSA Schedule task order/Blanket purchase agreement	May 15, 2016
Acquisition Name	Description	Number of Contracts	Procurement Vehicle	Base Award Date
Public Assistance Technical Assistance Contract III (PA-TAC)	Supports FEMA's objectives of providing public assistance in a timely, effective, and consistent manner.	4	Indefinite delivery contract	February 17, 2012
Individual Assistance Support Contract (IASC)	Supports individual assistance and mass care in zone 2—FEMA regions 6 through 10.	2	Indefinite delivery contract	March 8, 2018
Housing Inspection Services	Provides damage inspection services in affected disaster areas in support of claims for damaged homes.	2	Indefinite delivery contract	September 30, 2013
Mail Operations	Processes incoming and outgoing mail communication and claims from disaster victims.	2	Indefinite delivery contract	September 12, 2015
Crisis Counseling Program	Provides crisis counseling assistance and training.	1	Interagency reimbursable work agreement	October 1, 2017
Disaster Readiness Account – Disaster Unemployment Assistance	Provides unemployment assistance in times of disaster.	1	Interagency agreement	December 24, 2014

Acquisition Name	Description	Number of Contracts	Procurement Vehicle	Base Award Date
Missing and Exploited Child Alert Services	Provides alert services for missing children during response and recovery of federally declared disasters.	1	Indefinite delivery contract	March 22, 2018
Individual Identity Verification Service	Verifies and authenticates the identity of personnel applying for disaster assistance to prevent duplicate or fraudulent applications for assistance.	1	Indefinite delivery contract	September 30, 2013
National and Regional Commercial Ambulance Gulf/East Coast States	Provides air and ground ambulance support services in Atlantic and Gulf Coast states. Pacific and Central Coast states are non-priced optional line items.	1	Indefinite delivery contract	July 17, 2017
Aviation and Ground Support Services for Large Scale Disasters	Provides aviation and ground support evacuation at airports in Southeastern and Gulf Coast states.	1	Indefinite delivery contract	September 21, 2016
Executive Aircraft Transport	Agreement with the Federal Aviation Administration to use executive air travel during response and recovery efforts for federally declared disasters.	1	Interagency agreement	August 11, 2016
Air Transportation Support Services	Provides air transportation support services.	2	Indefinite delivery contract	April 12, 2017
Motor Coach Evacuation – Mass Transportation Services	Provides preparedness planning and operation support for large-scale motor coach evacuation of general population within the continental U.S.	1	Indefinite delivery contract	May 15, 2014

Table 3. (Continued)

Acquisition Name	Description	Number of Contracts	Procurement Vehicle	Base Award Date
Telephone Interpretation Services	Provides telephone interpretation services at National Processing Service Center in Denton, TX during a federally declared disaster.	2	Indefinite delivery contract	February 23, 2012
Rad Responder Support Services	Provides standardized radiation data collection, management, and monitoring of environmental data during a radiological or nuclear event.	1	GSA Schedule task order	August 8, 2016
Public Service Advertising Ad Council	Provides public alerts and campaigns to the general public prior to, during, and post-disaster.	1	Indefinite delivery contract	September 9, 2014
Mobile Disaster Recovery Center Satellite Airtime Service	Provides Mobile Disaster Recovery Center with satellite airtime communication and access to FEMA networks and communications for disaster operations.	1	Interagency agreement	December 21, 2014
National Responder Support Camp IT Services	Provides on-site IT support technicians, managed services, and satellite equipment.	1	GSA Schedule task order / Blanket purchase agreement	May 31, 2016
GPS Managed Services	Deploys contractors to the field to provide hands on GPS oversight.	2	Indefinite delivery contract	September 30, 2014
Deployment Tracking Systems	Tracks and deploys the national disaster workforce during response to federally declared disasters.	1	GSA Schedule task order	September 30, 2013 and September 30, 2015
Mobile Emergency Response Support Mobile Satellite Services	Provides mobile satellite services.	1	GSA Schedule task order	January 14, 2016

Acquisition Name	Description	Number of Contracts	Procurement Vehicle	Base Award Date
FEMA-Wide Wireless Telecommunications	Provides wireless telecommunications services to emergency responders.	2	Indefinite delivery contract	October 6, 2016
FEMA Wireline Telephone Services	Provides wireline and other telephone services for continental U.S. and outside continental U.S. operations.	1	Indefinite delivery contract	February 1, 2017
Centralized Operations, Maintenance and Management Information Technology	Provides operation and maintenance support for FEMA-wide systems.	4	Indefinite delivery contract	October 17, 2014
Logistics Supply Chain Management System	Provides contractor services to support operation and maintenance of Logistics Supply Chain Management System.	2	Contract(s)	August 2, and March 8, 2017
National Log House Program	Supports survivors during federally declared disasters.	6	Indefinite delivery contract	April 3, 2018
National Responder Support Camp Operation Services	Provides support services for responders, including mobilization and demobilization, site preparation, installation, and management of responder base camp.	6	Indefinite delivery contract	September 26, 2017
Consumable Medical Supply Shelter Support Items	Provides medical supply kit items for purchase on an as needed basis.	1	Indefinite delivery contract	September 26, 2017
Infant and Toddler Shelter Supply Items	Provides shelter supplies for children under the age of three.	1	Indefinite delivery contract	September 26, 2016
Logistics Support Commodities – Supplies and Services	Provides commodities and services such as tarps, cots, joint field office kits, pet supplies, and hygiene kits.	1	Interagency agreement	July 26, 2017

Table 3. (Continued)

Acquisition Name	Description	Number of Contracts	Procurement Vehicle	Base Award Date
Caribbean Deep Sea Freight	Provides deep see freight services to be activated by FEMA Region II when a storm is about to hit the U.S. Virgin Islands or Puerto Rico.	1	Indefinite delivery contract	August 31, 2016
Shelter Leased Generators and Equipment	Provides leased generators in support of shelter operations during federally declared disasters.	2	Indefinite delivery contract	March 14, 2015
Generator Maintenance – Continental U.S.	Provides maintenance to repair non-performing generator assets for distribution centers.	1	Contract	August 25, 2017
Generator Maintenance – Caribbean Distribution Center	Provides on-site generator maintenance for Caribbean distribution center.	1	Contract	September 30, 2017
Generator Maintenance – Hawaii and Guam Distribution Centers	Provides on-site generator maintenance for Hawaii and Guam distribution centers.	1	Contract	July 29, 2015
Durable Medical Equipment Cache	Provides durable medical equipment for purchase on an as needed basis.	1	Indefinite delivery contract	September 28, 2017
Commercial Cots	Provides commercial cots for continental U.S. and outside continental U.S. distribution centers.	1	Indefinite delivery contract	July 11, 2016
Supplies and Services with Defense Logistics Agency	Provides disaster commodities and services such as meals ready to eat, aviation, maritime, and land weapon systems, clothing, construction equipment, fuel, and	1	Interagency agreement	March 12, 2017

Acquisition Name	Description	Number of Contracts	Procurement Vehicle	Base Award Date
	medical supplies.			
Commercial and Reduced Sodium Meals	Provides commercial and reduced sodium meals for survivors.	3	Indefinite delivery contract	September 28, 2017
Manufactured Housing Units	Provides 1, 2, and 3 bedroom manufactured homes in support of disaster survivors.	7	Indefinite delivery contract	September 29, 2014
National Vehicle Maintenance Service Contract	Provides maintenance of disaster response tactical vehicles.	1	Indefinite delivery contract	May 29, 2018
Self-Help Tarps	Provides 20 by 25 foot self-help tarps for small areas of roof damage.	3	Indefinite delivery contract	September 12, 2014
Boxed and Bottled Water	Provides boxed and bottled water for disaster survivors.	4	Indefinite delivery contract	September 5, 2017
Blankets	Provides blankets for disaster survivors.	2	Indefinite delivery contract	September 27, 2013
Cleaning Supplies	Provides cleaning supplies to support survivors in shelters during federally declared disasters.	3	Indefinite delivery contract	September 27, 2013
Cross Docking Operations	Provides a network of transportation and emergency cross dock and freight forwarding service operation centers to support large scale trailer operations in the continental U.S.	1	Indefinite delivery contract	September 30, 2016
Miscellaneous Shelter Items	Provides miscellaneous shelter items to disaster survivors.	3	Indefinite delivery contract	September 26, 2013

Table 3. (Continued)

Acquisition Name	Description	Number of Contracts	Procurement Vehicle	Base Award Date
Shelter Clothing	Provides clothing for disaster survivors in shelters.	1	Indefinite delivery contract	August 2, 2016
Portable Facilities	Provides portable toilet facilities and fresh water delivery and removal services.	1	Indefinite delivery contract	September 16, 2016
Laundry Services	Provides laundry services for disaster survivors.	4	Indefinite delivery contract	May 30, 2017
Hazard Mitigation Technical Assistance Program Non- Architecture and Engineering	Provide non-architect and engineering technical assistance, research and analysis, and document development.	3	Indefinite delivery contract	March 20, 2015
Hazard Mitigation Technical Assistance Program Architect and Engineering	Provides production and technical services for architect and engineering contracts.	2	Indefinite delivery contract	November 24, 2014
Shipping and Receiving of Mail Equipment, Storage Services	Provides storage, shipping, and maintenance of mail equipment in support of disaster operations.	1	Purchase order	November 27, 2013
Warehouse Support Services	Provides warehouse support services.	1	Contract	May 4, 2015
Health and Human Services/ Federal Occupational Health Awareness Services	Provides environmental support, industrial hygiene and occupational safety	1	Interagency agreement	September 28, 2015

Acquisition Name	Description	Number of Contracts	Procurement Vehicle	Base Award Date
	support, hazardous waste, disaster safety, bio hazard safety support, and fire program development and support.			

Source: Fiscal Year 2018 FEMA Prepositioned Contract List, June 1, 2018. I GAO-19-93.

Note: Contracts listed represent all available FEMA advance contracts, as identified by their June fiscal year 2018 prepositioned contract list, and not just contracts used in response to the three 2017 hurricanes and California wildfires.

Table 4. U.S. Army Corps of Engineers (USACE) Identified Advance Contracts, as of September 2018

Advance Contract Initiative	Description	Number of Contracts	Procurement Vehicle	Base Award Year
Temporary Power	Provides emergency power to critical public facilities	4	Indefinite Delivery Contract	2013/2014
Temporary Roofing	Provides installation of blue plastic sheeting onto roofs of damaged homes or public structures	12	Indefinite Delivery Contract	2014/2015
Debris Removal	Provides emergency debris clearance and removal	17	Indefinite Delivery Contract	2014

Source: GAO review of USACE information. I GAO 19-93.

Note: Contracts listed represent all available USACE advance contracts, and not just advance contracts used in response to the three 2017 hurricanes and California wildfires.

APPENDIX IV: COMMENTS FROM THE DEPARTMENT OF HOMELAND SECURITY

U.S. Department of Homeland Security
Washington, DC 20528

Homeland Security

November 20, 2018

Marie A. Mak
Director, Contracting and National Security Acquisition
U.S. Government Accountability Office
441 G Street, NW
Washington, DC 20548

Re: Management Response to Draft Report GAO-19-93, "2017 DISASTER CONTRACTING: Action Needed to Better Ensure More Effective Use and Management of Advance Contracts" (Job Code 102660)

Dear Ms. Mak,

Thank you for the opportunity to review and comment on the subject draft report. The U.S. Department of Homeland Security (DHS) appreciates the U.S. Government Accountability Office's (GAO) work in planning and conducting its review and issuing this report.

The Department is pleased to note GAO's recognition of the positive steps FEMA has taken to institute training and develop guidance on timeframes related to the acquisition planning process for advance contracts. Contracting during disasters can pose a unique set of challenges and significant amount of pressure to provide life-sustaining goods and services to survivors as quickly as possible. Advance contracts significantly contribute to FEMA's ability to rapidly leverage cost-effective method for mobilizing resources, while also helping to preclude the need to procure critical goods and services non-competitively. FEMA remains committed to (1) contracting in full compliance with the Post-Katrina Emergency Management Reform Act (PKEMRA), Robert T. Stafford Disaster Relief and Emergency Assistance Act, Federal Acquisition Regulation (FAR), Homeland Security Acquisition Regulation, and Homeland Security Acquisition Manual, and (2) continuously communicating to state and local governments on the availability and proper establishment and use of the advance contracts.

The draft report contained nine recommendations with which the Department concurs. Attached for our detailed responses to each recommendation. Technical comments were previously provided under separate cover.

Again, thank you for the opportunity to review and comment on this draft report. Please feel free to contact me if you have any questions. We look forward to working with you again in the future.

Sincerely,

JIM H. CRUMPACKER, CIA, CFE
Director
Departmental GAO-OIG Liaison Office

Attachment: Management Response to Recommendations Contained in GAO-19-93

GAO recommended that FEMA's Administrator:

Recommendation 1: Update the strategy identified in its 2007 *Advance Contracting of Goods and Services Report to Congress* to clearly define the objectives of advanced contracts, how they contribute to FEMA's disaster response operations, and whether and how they should be prioritized in relation to new post-disaster contract awards.

Response: Concur. FEMA's Office of the Chief Component Procurement Officer (OCPO) will update the strategy identified in the 2007 Advance Contracting of Goods and Services Report to Congress in the *Disaster Contracting Desk Guide*, outlining clear objectives and use of advance contracts in disaster operations. FEMA will memorialize a structure that prioritizes utilizing existing advance contracts related to new-post disaster contract awards. Requirements for goods and services determined to meet the requirement of advance contracts, in accordance with PKEMRA and Robert T. Stafford Disaster Relief and Emergency Assistance Act, are identified in the 5-year Master Acquisition Planning Schedule (MAPS) for accountability and advance acquisition planning purposes. Estimated Completion Date (ECD): June 30, 2019.

Recommendation 2: Ensure the Chief Procurement Officer updates the *Disaster Contracting Desk Guide* to include guidance for whether and under what circumstances contracting officers should consider using existing advance contracts prior to making new post-disaster contract awards, and include this guidance in existing semi-annual training given to the contracting officers.

Response: Concur. FEMA OCPO will update the current *Disaster Contracting Desk Guide* to establish clear guidance prioritizing the use of existing advance contracts prior to issuing any new post-disaster contract awards. The updated guidance will be communicated to contracting personnel during the next disaster webinar. ECD: June 30, 2019.

Recommendation 3: Update and implement existing guidance for program office and contracting officer personnel to identify acquisition planning timeframes and considerations across the entire acquisition process and clearly communicate the purpose and use of MAPS.

Response: Concur. FEMA's OCPO conducted outreach with all of the program offices during June 2016 on the purpose and use of the 5-year MAPS program. This ensured all stakeholders have complete understanding of the tool. As a result, the program offices have full visibility into future procurement actions, and work with OCPO to conduct acquisition planning within established timelines. As a part of these efforts, all program offices receive communication from FEMA's OCPO 18 months before the expiration of their contract. This communication requests a program manager to work with OCPO Portfolio Managers to ensure solicitations are prepared in a timely manner, ensuring timely award. We request that GAO consider this recommendation resolved and closed as implemented.

Recommendation 4: Ensure the Chief Procurement Officer adheres to current hard copy contract file management requirement to ensure advance contract files are complete and up to date, whether they will be transferred into the new Electronic Contract Filing System [ECFS] or remain in hard copy format.

Response: Concur. FEMA OCPO will update the current Labeling and Organizing Official Contract File Folders Standard Operating Procedure (SOP), dated August 31, 2009, to clearly institute and enforce the policy of maintaining the official record of contract file as a hard copy unless rescinded with the implementation of a Department-wide ECFS. ECD: June 30, 2019.

Recommendation 5: Ensure the Chief Procurement Officer reminds contracting officers of the three day timeframe for entering completed award documentation into contracting writing system when executing notice to proceed documentation.

Response: Concur. FEMA OCPO will update the current Notice To Proceed SOP, dated April 23, 2013, to ensure that contracting officers know all awards must be entered into contracting writing system and Federal Procurement Data System (FPDS) within three days of award. ECD: July 31, 2019.

Recommendation 6: Ensure the Chief Procurement Officer revises its reporting methodology to ensure that all disaster contracts are included in its quarterly reports to congressional committees on disaster contract actions.

Response: Concur. To ensure accurate reporting of all disaster contracts to congressional committees. FEMA OCPO will incorporate an addendum in its quarterly report. The addendum will capture the contract actions that were not previously reported based on the exemption of FAR 4.604(b)(3), which allows any action awarded in accordance with FAR 6.302-2 or pursuant to any of the authorities listed at FAR Subpart 18.2 to be submitted in FPDS within 30 days after contract award. ECD: July 31, 2019.

Recommendation 7: Ensure the Chief Procurement Officer revises the *Disaster Contracting Officer Desk Guide* to provide specific guidance for contracting officer to perform outreach to state and local governments on the use and establishment of advance contracts.

Response: Concur. FEMA OCPO has taken the initiative to create an external acquisition toolbox website for the state, local, territorial and tribal (SLTT) governments to access templates and tools needed to establish SLTT governments' own advance contracts. Additionally, a procurement hotline was established August 23, 2018 to answer any questions or provide additional assistance when needed/requested by the SLTT governments. Finally, FEMA OCPO will update the current *Disaster Contracting Desk Guide* to provide clear and specific guidance on coordination and outreach to governments. ECD: October 31, 2019.

Recommendation 8: Ensure the Chief Procurement Officer identifies a single centralized resource listing its advance contracts and ensure that source is updated regularly to include all available advance contracts.

Response: Concur. A list of all advance contracts is provided to contracting personnel and external customers through the OCPO Reporting Dashboard SharePoint site. The list is also updated monthly and distributed to all OCPO supervisors on a monthly basis, and is available on FEMA's intranet website. We request that GAO consider this recommendation resolved and closed as implemented.

Recommendation 9: Ensure the Chief Procurement Officer communicates information on available advance contracts through the centralized resource to states and localities to inform their advance contracting efforts.

Response: Concur. FEMA's OCPO, in coordination with the Office of External Affairs and Procurement Disaster Assistance Team (PDAT) within FEMA Office of Chief Counsel, will conduct outreach with SLTT governments to communicate establishing the acquisition toolbox website. The toolbox will provide SLTTs access to the list of advance contracts, templates, and tools needed to establish their own advance contracts. In addition, PDAT will continue to engage SLTT governments and communicate guidance on procurement under grants through in-person and virtual trainings, deployments, technical assistance, and its resources and tools on the PDAT website. ECD: October 31, 2019.

In: Hurricanes, Wildfires and Flooding ISBN: 978-1-53616-362-9
Editor: Lydie Yohan

Chapter 11

2017 DISASTER CONTRACTING: ACTIONS NEEDED TO IMPROVE THE USE OF POST-DISASTER CONTRACTS TO SUPPORT RESPONSE AND RECOVERY*

United States Government Accountability Office

ABBREVIATIONS

DART	Disaster Acquisition Response Team
DHS	Department of Homeland Security
DLA	Defense Logistics Agency
DOD	Department of Defense
FAR	Federal Acquisition Regulation
FEMA	Federal Emergency Management Agency
FPDS-NG	Federal Procurement Data System – Next Generation

* This is an edited, reformatted and augmented version of United States Government Accountability Office; Report to Congressional Requesters, Accessible Version, Publication No. GAO-19-281, dated April 2019.

GSA	General Services Administration
NIA	National Interest Action
OCPO	Office of the Chief Procurement Officer
PKEMRA	Post-Katrina Emergency Management Reform Act
USACE	U.S. Army Corps of Engineers

Why GAO Did This Study

Federal contracts play a key role in timely response and recovery efforts following disasters. While federal agencies, such as FEMA and USACE, may have advance contracts in place for obtaining goods and services following disasters, agencies may also award post-disaster contracts.

GAO was asked to review the federal government's response to three major hurricanes in 2017, as well as the 2017 California wildfires. This chapter addresses, among other objectives, the extent to which (1) federal agencies obligated funds on post-disaster contracts in response to the these events, and (2) selected agencies experienced challenges in the planning of selected contracts.

GAO analyzed data from the Federal Procurement Data System-Next Generation; selected a non-generalizable sample of 23 post-disaster contracts based on factors such as if the contract was set aside for award to a local contractor; reviewed federal regulations and agency guidance; and interviewed agency officials.

What GAO Recommends

GAO is making 10 recommendations, including that DHS reopen NIA codes for Hurricanes Harvey and Irma; USACE provide guidance on the local area definition; and the Coast Guard and USACE provide guidance to ensure contracting requirements for the use of non-local vendors are met.

Agencies concurred with 9 recommendations. DHS did not agree that NIA codes should be reopened. GAO continues to believe DHS should do so, to the extent practicable, as discussed in the report.

WHAT GAO FOUND

Following hurricanes Harvey, Irma, and Maria and the 2017 California wildfires, federal agencies obligated at least $5 billion in post-disaster contracts—which are awarded after disasters hit— to support disaster response and recovery efforts. The U.S. Army Corps of Engineers (USACE) and the Federal Emergency Management Agency (FEMA) comprised over three-quarters of reported post-disaster contract obligations as of June 30, 2018 (see Figure).

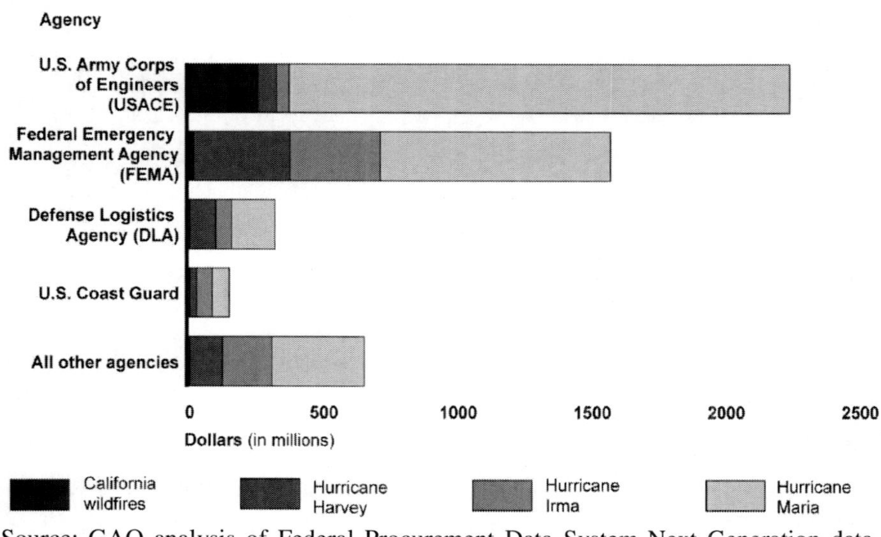

Source: GAO analysis of Federal Procurement Data System-Next Generation data. | GAO-19-281.

Known Post-Disaster Contract Obligations in Response to the 2017 Disasters, in Fiscal Year 2018 Dollars, as of June 30, 2018.

However, the full extent of post-disaster contracting related to the 2017 disasters is unknown due to the Department of Homeland Security's (DHS) inconsistent implementation of the criteria for closing a national interest action (NIA) code. This code allows agencies to track data on contract actions related to national emergencies, providing government-wide insight into response and recovery efforts. DHS closed the codes for Harvey and Irma on June 30, 2018, less than a year after those hurricanes hit. In contrast, the codes for prior hurricanes were open for at least five years, with Katrina remaining open for 13 years.

Based on a review of 23 contract files from FEMA, USACE, the Defense Logistics Agency, and the Coast Guard, GAO identified challenges in the planning of selected contracts. For example, GAO found USACE officials were not consistently aware of the regulation that defines "local area." GAO also found that contracting officers at FEMA, USACE, and the Coast Guard did not consistently write justifications for awards to non-local vendors outside the disaster area, as required. FEMA developed guidance to address this, but the Coast Guard and USACE have not issued guidance or tools to address this requirement. Without addressing planning challenges, agencies may miss opportunities to award contracts to local businesses in the disaster area to the extent feasible and practicable, which could help jump-start the local economy.

April 24, 2019
Congressional Requesters

Federal contracts for life-saving or life-sustaining goods and services play a key role in the immediate aftermath of a disaster and in long-term community recovery. In 2017, four disasters—hurricanes Harvey, Irma, and Maria, and the California wildfires—occurred nearly back-to-back and collectively affected 47 million people, or about 15 percent of the nation's population, according to the Federal Emergency Management Agency (FEMA), a component of the Department of Homeland Security (DHS). Federal agencies obligated billions of dollars on contracts in response to those disasters. Our prior work identified issues related to post-disaster contracts—which are awarded after a disaster hits. For example, in September 2015, we found that FEMA's contracting officers displayed limited awareness of requirements enacted after Hurricane Katrina to provide a contracting preference to local firms for disaster response contracts to the extent feasible and practicable and to limit the length of noncompetitive contracts awarded based on the urgency exception.[1]

You asked us to review the federal government's contracting efforts for preparedness, response, and recovery efforts related to the three 2017 hurricanes and California wildfires. This chapter specifically addresses the use of post-disaster contracts and: (1) assesses the extent to which federal agencies obligated funds on post-disaster contracts in response to the 2017 major disasters; (2) assesses the extent to which selected agencies experienced challenges in the planning process for selected post-disaster contracts; and (3) describes selected agencies' lessons learned as a result of the 2017 major disasters and assesses the extent to which they have taken action to address them. In December 2018, we reviewed and reported on these areas on the use of advance contracts, which are contracts for life-sustaining goods and services that are set up prior to disasters to be used in the immediate aftermath of disasters.[2]

To identify the extent to which federal agencies obligated funds on post-disaster contracts in response to the 2017 disasters, we reviewed Federal Procurement Data System-Next Generation (FPDS-NG) data through June 30, 2018, the most recent and complete data available at the time of our review.[3] We identified obligations related to Hurricanes

---

[1] GAO, *Disaster Contracting: FEMA Needs to Cohesively Manage Its Workforce and Fully Address Post-Katrina Reforms*, GAO-15-783 (Washington, D.C.: Sept. 29, 2015).

[2] GAO, 2017 *Disaster Contracting: Action Needed to Better Ensure More Effective Use and Management of Advance Contracts,* GAO-19-93 (Washington, D.C.: Dec. 6, 2018).

[3] We adjusted the obligation data to constant fiscal year 2018 dollars using the Fiscal Year Gross Domestic Product price index. For the purposes of this chapter, contract obligations include obligations against what the General Services Administration's FPDS-NG categorizes as definitive vehicles (definitive contracts and purchase orders that have a defined scope of work that do not allow for individual orders under them), and against what FPDS-NG categorizes as indefinite delivery vehicles (orders under the Federal Supply Schedule, orders/calls under blanket purchase agreements, orders under basic ordering agreements, orders under government-wide acquisition contracts, and orders under other indefinite delivery vehicles, such as indefinite delivery, indefinite quantity contracts). DHS and the Department of Defense (DOD) exercised the use of the special emergency procurement authorities within 41 U.S.C § 1903 (and as implemented by DHS FAR Class Deviation 17-02 and DOD FAR Class Deviation 2017-O0007, respectively) to increase the micro-purchase threshold to $20,000 for procurements in support of these major disaster

Harvey, Irma, and Maria using the national interest action (NIA) code, as well as the contract description. NIA codes are a data field established in FPDS-NG for the purpose of tracking federal procurement related to a specific disaster, emergency, or contingency event. We also assessed the process for establishing and closing a NIA code. Specifically, we reviewed the criteria in the 2012 and 2018 memorandum of agreement between DHS, the Department of Defense (DOD), and the General Services Administration (GSA)—which oversees FPDS-NG—and interviewed officials involved in the process. Because a NIA code was not established in FPDS-NG for the 2017 California wildfires, we asked the agencies with the highest obligations on post disaster contracts for the three hurricanes—FEMA, U.S. Army Corps of Engineers (USACE), Defense Logistics Agency (DLA), and the U.S. Coast Guard (Coast Guard)—to identify contracts that they used to respond to that event.[4] Therefore, our analysis does not capture whether other agencies' obligated funds on contracts related to the 2017 California wildfires. We assessed the reliability of FPDS-NG data by reviewing existing information about the FPDS-NG system and the data it collects— specifically, the data dictionary and data validation rules—and performing electronic testing. We also compared FPDS-NG data to the contract files in our review. We determined the FPDS-NG data were sufficiently reliable for the purposes of describing agencies' reported obligations on post-disaster contracts.

To assess the extent to which selected agencies experienced challenges in the planning process—such as recognizing the requirements of contracting with local vendors—for selected post-disaster contracts, we reviewed relevant laws and regulations, including the Post-Katrina Emergency Management Reform Act (PKEMRA), the Federal Acquisition

---

responses. As a result, contract obligations for hurricanes reported in FPDS-NG may only include obligations over that amount.

[4] Obligations on post-disaster contracts for the California wildfires reflect agency identified obligations for Disaster Response 4344 and Disaster Response 4353. Coast Guard officials stated that they did not award any contracts in response to the 2017 California wildfires. DLA officials stated that they have some contracts in place which for the most part, provide inventory replenishment for DLA and the U.S. Forest Service within the U.S. Department of Agriculture. However DLA was unable to provide data on contracts awarded specifically for response and recovery related to the two wildfire disasters in the scope of our review.

Regulation (FAR), and the Robert T. Stafford Disaster Relief and Emergency Assistance Act (Stafford Act), as well as agency policy and guidance. We identified a non-generalizable sample of 23 post-disaster contracts from the four agencies with the highest post-disaster obligations—FEMA, USACE, DLA, and Coast Guard—based on FPDS-NG data as of March 31, 2018. We selected contracts to obtain a range of contracts across four primary selection criteria: (1) contracts awarded using the unusual and compelling urgency exception to full and open competition ("urgency exception"); (2) contracts awarded using a local area set-aside (wherein only vendors residing or primarily doing business in the declared disaster area may compete for a contract award); (3) contracts awarded to small businesses; and (4) contracts terminated for default or convenience.[5] We also selected contracts across all three hurricanes and the California wildfires and all four of the selected agencies. Using these criteria, we selected 12 FEMA, seven USACE, two DLA, and two Coast Guard contracts (see Table 1).

We assessed planning documentation in these contracts against applicable regulatory and statutory criteria. For example, we assessed the eight local area set-aside contracts against the related FAR and Stafford Act requirements, and we assessed the 11 contracts using the urgency exception against the relevant FAR and PKEMRA criteria. We met with officials from FEMA, USACE, DLA, the Coast Guard, and the Office of Management and Budget's Office of Federal Procurement Policy. Findings based on information collected from the 23 contracts cannot be generalized to all post-disaster contracts.

[5] Termination for convenience is the government's unilateral contractual right to partially or completely terminate a contract, when it is in the government's interest. For example, a contracting officer may terminate for convenience when the requirement is no longer needed. When terminating for convenience the government is required to come to a settlement that compensates the contractor for the work done and preparations made for the terminated portion of the contract. The government may terminate all or a portion of a commercial item contract for cause if the contractor fails to provide items or services that conform to the contract requirements. A termination for default is generally the exercise of the Government's contractual right to completely or partially terminate a contract because of the contractor's actual or anticipated failure to perform its contractual obligations.

Table 1. Contract File Selection by Selection Criteria

Incident	Urgency Exception Cited	Local Area Set-Aside	Small Business Vendor	Terminated Contract	Total Selected[a]
FEMA Hurricane Harvey	1	3	3	0	7
FEMA Hurricane Irma	0	1	1	0	2
FEMA Hurricane Maria	1	2	4	3	10
US. Army Corp Engineers California Wildfires	2	0	0	0	2
US. Army Corp Engineers Hurricane Maria[b]	3	2	3	0	8
Defense Logistics Agency Hurricane Maria	2	0	1	0	3
U.S. Coast Guard U.S. Coast Guard Hurricane Irma	1	0	0	0	1
Hurricane Maria[c]	1	0	1	0	2
Total Contracts	11	8	13	3	23

Source: Selected Contracts Based on GAO Analysis of Federal Procurement Data System-Next Generation Data | GAO-19-281.

[a]Numbers may not sum to total selected contracts column because some contracts met multiple selection criteria.

[b]One contract had a small amount of obligations for Hurricane Irma.

[c]This contract had a small amount of obligations for Hurricane Irma.

To describe what lessons learned selected agencies identified related to the use of post-disaster contracts and assess the extent to which agencies have taken action to address them, we reviewed available completed after-action reports for the 2017 disasters from agencies with the highest obligations on post-disaster contracts for the three hurricanes—FEMA, USACE, DLA, and the Coast Guard—and interviewed officials from these agencies.[6] We also reviewed interagency lessons learned from the Emergency Support Function Leadership Group—a body of senior officials from each of the national emergency support functions, along with FEMA headquarters and regional officials, tasked with coordinating

[6] After-Action Reports identify lessons learned and areas for improvement and may be completed following a training exercise or a real-world event.

responsibilities and resolving operational and preparedness issues relating to interagency response activities in support of the National Response Framework. To obtain perspectives and examples from state and local government officials involved in disaster response, we interviewed officials in California on the use of federal contracts. We also met with state and local officials in Texas, Florida, Puerto Rico, and the U.S. Virgin Islands to discuss the federal response to the 2017 hurricanes. The information gathered from these officials is not generalizable to all officials. Appendix I provides more information about our overall scope and methodology.

We conducted this performance audit from March 2018 to April 2019 in accordance with generally accepted government auditing standards. Those standards require that we plan and perform the audit to obtain sufficient, appropriate evidence to provide a reasonable basis for our findings and conclusions based on our audit objectives. We believe that the evidence obtained provides a reasonable basis for our findings and conclusions based on our audit objectives.

BACKGROUND

In 2017, three major hurricanes made landfall in the United States and historic wildfires struck California. According to FEMA, the 2017 hurricanes and wildfires collectively affected 47 million people—nearly 15 percent of the nation's population. See Figure 1 for a timeline of these major disasters.[7]

[7] A major disaster is any natural catastrophe (including any hurricane, tornado, storm, high water, wind driven water, tidal wave, tsunami, earthquake, volcanic eruption, landslide, mudslide, snowstorm, or drought), or, regardless of cause, any fire, flood, or explosion, in any part of the United States, which the President determines causes damage of sufficient severity and magnitude to warrant major disaster assistance to supplement the efforts and available resources of states, local governments, and disaster relief organizations in alleviating damage, loss, hardship, or suffering. See 42 U.S.C. § 5122(2).

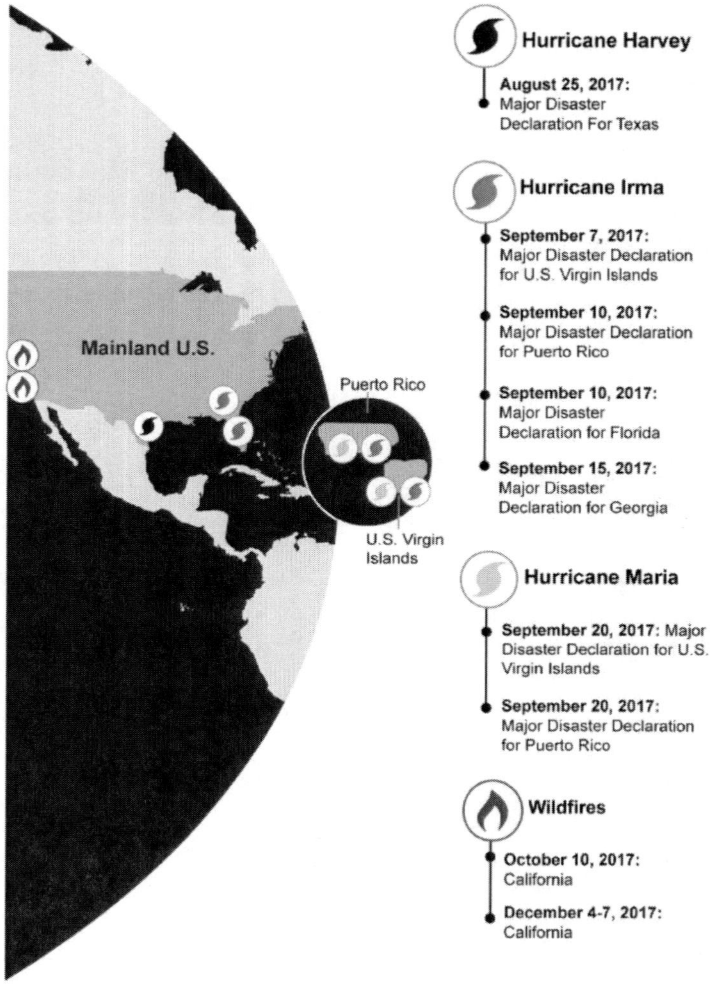

Source: GAO analysis of Federal Emergency Management Agency data. | GAO-19-281.

Figure 1. Timeline of Major 2017 Hurricanes and California Wildfires.

Overview of Federal Disaster Response and Recovery

When disasters hit, state and local entities are typically responsible for disaster response efforts. The Stafford Act establishes a process by which the Governor of the affected state or the Chief Executive of an affected Indian tribal government may request a presidential major disaster declaration to obtain federal assistance.[8] According to the DHS National Response Framework—a guide to how the federal government, states and localities, and other public and private sector institutions should respond to disasters and emergencies—the Secretary of Homeland Security is responsible for ensuring that federal preparedness actions are coordinated to prevent gaps in the federal government's efforts to respond to all major disasters, among other emergencies.[9] The framework also designates FEMA as the lead agency to coordinate the federal disaster response efforts across 30 federal agencies.

The National Response Framework identifies 14 emergency support functions that serve as the federal government's primary coordinating structure for building, sustaining, and delivering disaster response efforts across more than 30 federal agencies. Each function defines specific mission areas—such as communication, transportation, and energy—and designates a federal department or agency as the coordinating agency. For example, provision of assets and services related to public works and engineering, such as temporary roofing or power, are coordinated by USACE within DOD.[10] See Appendix II for more information about emergency support function responsibilities across the federal government.

FEMA's Response Directorate coordinates disaster response efforts through mission assignments—work orders that it issues to other federal agencies to direct them to utilize their authorities and the resources granted

---

[8] 42 U.S.C. § 5170. The Robert T. Stafford Disaster Relief and Emergency Assistance Act, as amended, permits the President to declare a major disaster after a state's governor or chief executive of an affected Indian tribal government—a governing body of an Indian or Alaska Native tribe, band, nation, pueblo, village, or community that is federally recognized—finds that the emergency or major disaster is of such a severity and magnitude beyond the State, Indian tribal government, and local government's capabilities. Governor means the chief executive of any state, which includes, among others, Puerto Rico and the U.S. Virgin Islands. 42 U.S.C.§ 5122 (4)(5).

[9] Department of Homeland Security, *National Response Framework* (June 2016).

[10] USACE has both military and civilian responsibilities. The Chief of Engineers, a military officer, oversees USACE's civil works operations and reports on civil works matters to the Assistant Secretary of the Army for Civil Works, who establishes the policy direction for the civil works program.

to them under federal law in support of direct assistance to state, local, tribal, and territorial governments.

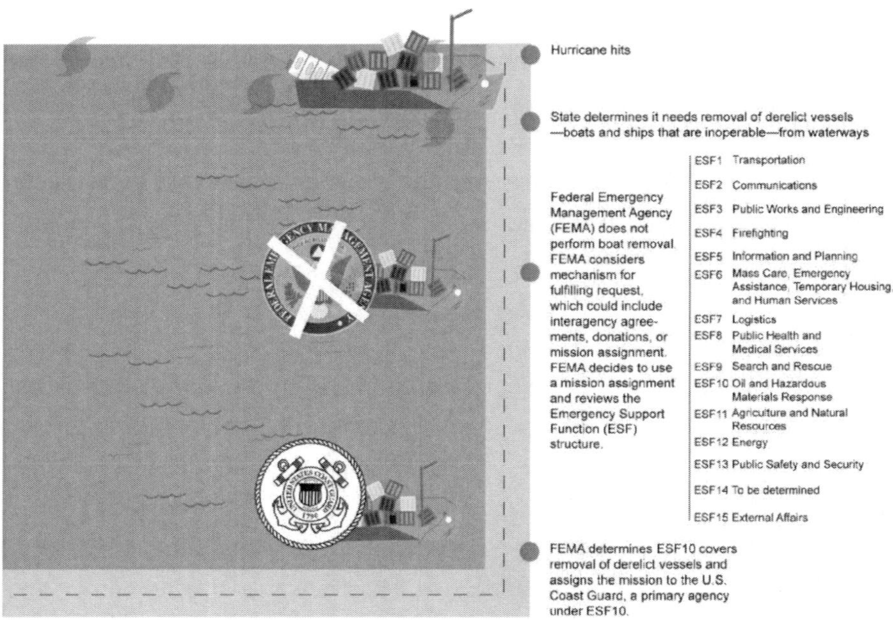

Source: GAO analysis of FEMA mission assignment process. | GAO-19-281.

Note: Emergency support function 14 is in transition to the Cross-sector Business and Infrastructure Coordination function.

Figure 2. Notional Mission Assignment Process for Removal of Derelict Marine Vessels.

Mission assignments are authorized by the Stafford Act, and agencies may fulfill these assignments through federal contracts. FEMA made 1,515 mission assignments for the 2017 hurricanes and California wildfires, and total obligations for these mission assignments were more than $7.8 billion as of January 2018, according to FEMA. See Figure 2 for a depiction of the mission assignment process under a notional scenario of removing derelict marine vessels—boats and ships damaged during a hurricane and that are determined to be inoperable.

The National Response Framework states that when an Emergency Support Function is activated in response to an incident, the primary agency for that emergency support function is responsible for executing contracts and procuring goods and services as needed, among other things. For example, DOD and USACE are the coordinators for Emergency Support Function 3—public works and engineering—and as part of this role, these agencies are responsible for emergency contracting support for lifesaving and life-sustaining services. As such, during the 2017 disasters, USACE obligated funds on contracts in support of its assigned mission of public works and engineering by restoring the electrical grid in Puerto Rico following Hurricane Maria and removing debris following the California wildfires.

FEMA's Contracting Workforce

In its role as the lead coordinator of federal disaster response efforts across federal agencies, FEMA's contracting workforce plays a key role in post-disaster contracts. FEMA's contracting efforts are supported by its contracting workforce within FEMA's Office of the Chief Procurement Officer (OCPO). In our prior work, we found that FEMA's contracting workforce had grown significantly since Hurricane Katrina, but the agency struggled with attrition at times.[11] While the majority of FEMA's contracting workforce is located in headquarters, contracting officers are also located in each of FEMA's 10 regional offices. See Figure 3 for the location of FEMA's 10 regional offices as well as the states and territories for which each one is responsible in terms of fulfilling National Response Framework duties.

In addition, FEMA can deploy members of its Disaster Acquisition Response Team (DART), a group whose primary purpose is to support contract administration for disasters.[12] There are two DART teams under FEMA's Expeditionary branch, each comprised of contracting officers,

[11] GAO-15-783.

[12] In 2010, FEMA created 18 new contracting officer positions to form DART. DART personnel are considered headquarters employees for management purposes, but can be based in FEMA headquarters in Washington, D.C. or in one of three regional offices (Atlanta, Georgia; Baton Rouge, Louisiana; or Oakland, California).

contracting specialists, and quality assurance specialists. Figure 4 shows how FEMA's contracting workforce is organized.

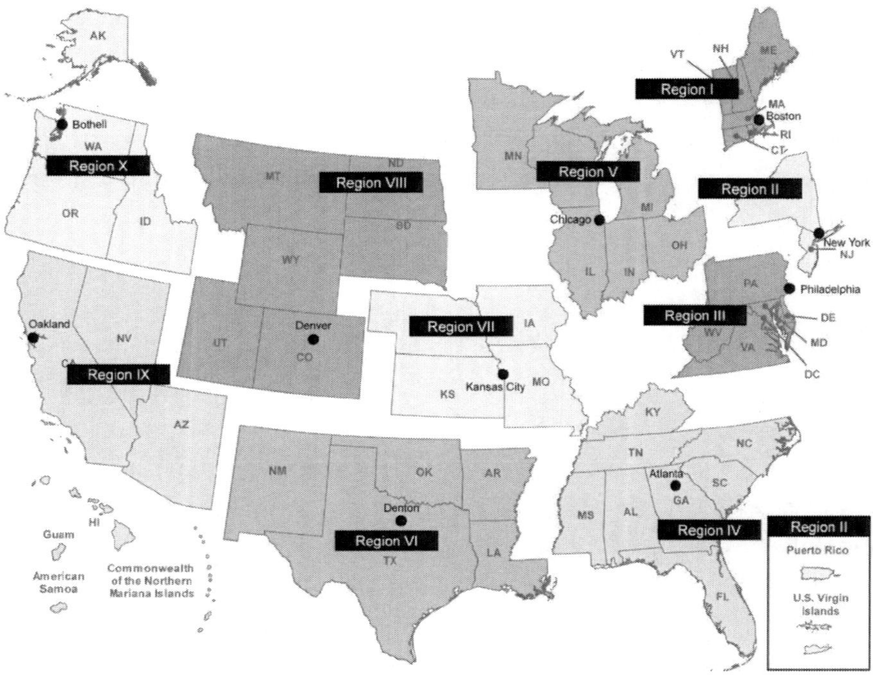

Source: Federal Emerging Management Agency; Map Resources (map). | GAO-19-281.

Figure 3. Map of Federal Emergency Management Agency (FEMA) Regions and Their Member States and Territories.

In headquarters, FEMA's contracting officers support a variety of functions, such as contracting for information technology needs, activities to prepare for and mitigate disasters, and disaster response. In the field, the disaster and field operations division manages contracting for disaster response efforts including:

- Logistics: delivering goods and services to support disaster survivors and communities, including life-sustaining commodities such as meals, blankets, and electricity generators,

- Response: coordinating capabilities needed immediately following a disaster, such as air and ground evacuation services and emergency sheltering, and
- Recovery: primarily supporting rebuilding efforts, including technical assistance programs.

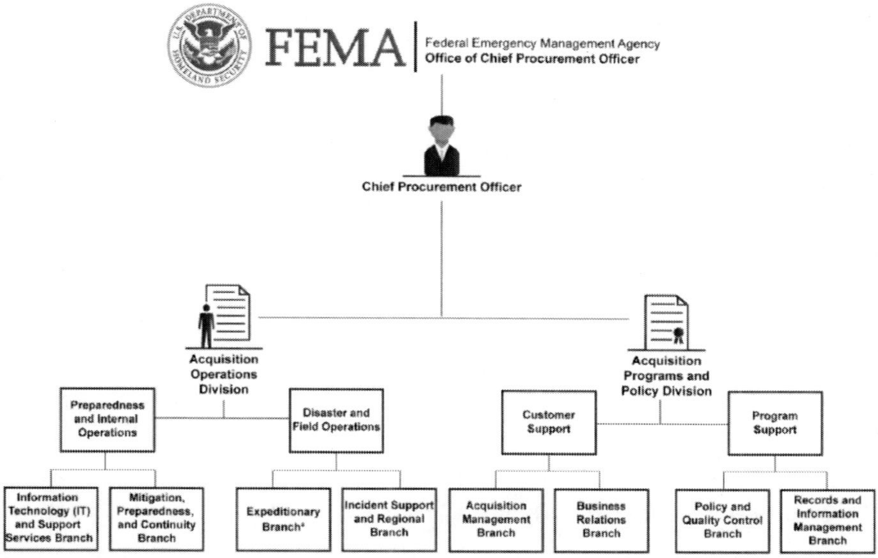

Source: GAO analysis of FEMA data. | GAO-19-281.

[a]The Expeditionary branch includes two Disaster Acquisition Response Teams.

Figure 4. FEMA Office of the Chief Procurement Officer.

Regional Contracting Officers

Regional contracting officers serve as the first response for contracting if a disaster occurs in their region. During a disaster, the regional offices can request additional contracting support from headquarters if needed. Contracting officers are typically located in each regional office's mission support division, which provide essential administrative, financial, information technology, and acquisition support for the region. Each region is headed by a Regional Administrator who reports directly to the head of FEMA, the FEMA Administrator.

In response to a 2009 DHS Inspector General Report, FEMA created a formal agreement to establish a new role for FEMA's OCPO to oversee regional contracting staff.[13] The Inspector General report found that regional contracting officers only reported to their respective supervisor in the region—who usually are not contracting officers—with no formal link to FEMA's OCPO. The Inspector General recommended that only contracting officials should manage the technical performance of contracting officers. The report stated that having the contracting officer's performance and career advancement controlled by someone who is not a contracting professional was an internal control risk and created a potential conflict-of-interest situation for the contracting officer. A subsequent 2011 agreement between the regions and headquarters states that a FEMA OCPO official will be the contracting officers' performance reviewer and that the regional supervisors will continue to manage regional contracting officials' day-to-day activities. As a result, regional contracting officers have a dual reporting chain to both FEMA OCPO in headquarters and to their supervisor within the region.

In September 2015, we identified challenges with how the agreement was being implemented, particularly in that it heightened the potential for an environment of competing interests for the regional contracting officers.[14] Specifically, we found that being physically located in a regional office where their regional supervisor is not a contracting professional gave contracting officers less standing to resist requests to perform duties outside of a contracting officer's responsibilities or to resist pressure from program officials to make certain decisions. Further, we found that FEMA had not updated its 2011 agreement, even though the agreement states that FEMA OCPO and the regions will revisit it each year. We recommended that the FEMA Administrator direct FEMA OCPO and the regional administrators to revisit the 2011 agreement to, among other things, add details about the extent of operational control headquarters and regional supervisors should exercise to minimize potential competing interests

---

[13] Department of Homeland Security Office of Inspector General, *Internal Controls in the FEMA Disaster Acquisition Process*, OIG-09-32 (Washington, D.C.: Feb. 19, 2009).

[14] GAO-15-783.

experienced by regional contracting officers, and further detail headquarters and regional supervisors' roles and responsibilities for managing regional contracting officers to improve coordination and communication. We also recommended, and FEMA agreed, that it establish a plan to review this agreement on an annual basis. As of January 2019, FEMA had not implemented these recommendations.

Joint Field Offices

After a major disaster is declared, FEMA establishes a joint field office, a temporary office through which it coordinates disaster response and recovery efforts with state and local governments and organizations. Once the need for disaster response and recovery ends and a joint field office is closed, the contracts supporting the disaster are returned to the cognizant regional contracting office.[15]

Post-Katrina Emergency Management Reform Act Contracting Requirements

Congress enacted the Post-Katrina Emergency Management Reform Act of 2006 (PKEMRA) after shortcomings were identified in preparation for and response to Hurricane Katrina—one of the largest and most destructive natural disasters in U.S. history, which hit the Gulf Coast in 2005.[16] PKEMRA included several provisions related to contracting, including:

- Contracting preference for local vendors. PKEMRA amended the Stafford Act to provide a contracting preference for local vendors.

[15] We are currently assessing the overall capacity, training, and qualifications of FEMA workforce to meet all disaster missions, including contracting and acquisition, during disaster deployments.

[16] PKEMRA was enacted as Title VI of the Department of Homeland Security Appropriations Act, 2007, Pub. L. No. 109-295, § 611, codified, as amended, at 42 U.S.C § 5150. The provisions of PKEMRA became effective upon enactment, October 4, 2006, with the exception of certain organizational changes related to FEMA, most of which took effect on March 31, 2007.

Specifically, for contracts or agreements with private entities, the provisions of the act state, in part:

- in general, for major disaster assistance activities, agencies shall provide a preference, to the extent feasible and practicable, to organizations, firms, and individuals residing or doing business primarily in the area affected by the major disaster or emergency;
- they may be set aside for local vendors, which means that only vendors residing or primarily doing business in the declared disaster area are allowed to compete for an award;
- those not awarded to local vendors shall be justified in writing in the contract file.

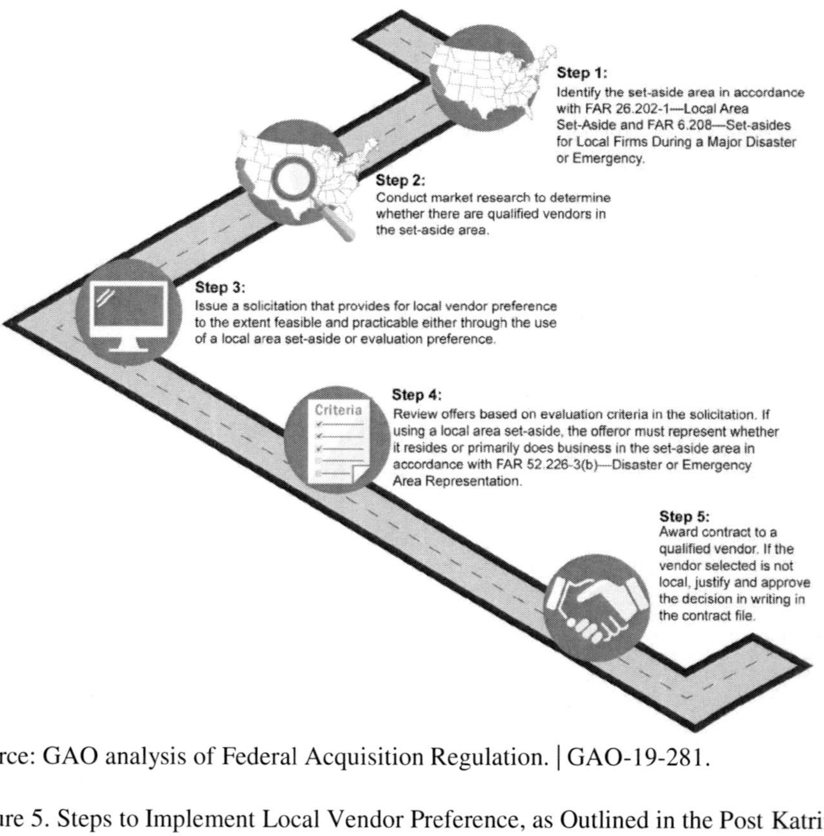

Source: GAO analysis of Federal Acquisition Regulation. | GAO-19-281.

Figure 5. Steps to Implement Local Vendor Preference, as Outlined in the Post Katrina Emergency Reform Act and the Federal Acquisition Regulation (FAR).

After the enactment of PKEMRA, changes were made to the FAR to implement provisions regarding the award of set-aside contracts to local vendors. Figure 5 displays the steps a contracting officer must take to implement the preference for awarding post-disaster contracts to a local vendor based on related laws and regulation.

- Use of noncompetitive contracts using the urgency exception. Agencies are generally required to use full and open competition—achieved when all responsible sources are permitted to compete—when awarding contracts. The Competition in Contracting Act of 1984 recognizes that full and open competition is not feasible in all circumstances and authorizes contracting without full and open competition under certain conditions, such as in cases with an unusual and compelling urgency and the government would be seriously injured unless the agency is permitted to limit the number of sources from which it solicits offers ("urgency exception").[17] When DHS awards disaster contracts non-competitively based on the urgency exception, PKEMRA, as implemented in the Homeland Security Acquisition Regulation, restricts the period of performance to 150 days, unless the Head of Contracting Activity determines that exceptional circumstances apply. For other uses of the urgency exception, the FAR's period of performance limit is generally no more than one year.[18] Generally, exceptions to full and open competition must be supported by written justifications that contain sufficient facts and rationale to justify use of the specific exception. Depending on the proposed value of the

[17] Pub. L. No. 98-369, §§ 2711(a)(1) and 2723(a)(1), codified, as amended, at 41 U.S.C. § 3301 and 10 U.S.C. § 2304. In all, there are seven exceptions to full and open competition. See FAR § 6.302.

[18] Under the FAR and DHS regulations (HSAR), the contract could exceed these time frames if the head of the agency determines exceptional circumstances apply. FAR § 6.302-2(d)(ii) and HSAR § 3006.302-270(d)(1)(iii).Under the FAR and HSAR, the period of performance of contracts using the urgency exception is limited to the time necessary to meet the unusual and compelling requirements and for the agency to enter into another contract for the requirements through the use of competitive procedures. FAR § 6.302- 2(d) and HSAR § 3006.302-270.

contract, the justifications require review and approval at successively higher approval levels within the agency.

- Use of advance contracts. PKEMRA requires FEMA to establish advance contracts, which are typically needed to quickly provide life-sustaining goods and services, such as tarps and meals, in the immediate aftermath of disasters. While not required under PKEMRA, USACE also establishes advance contracts for supplies and services (e.g., generators for its temporary power mission) using its independent statutory authorities for emergency management, such as Section 5 of the Flood Control Act of 1941.[19] In addition, DLA has an interagency agreement with FEMA to provide disaster commodities and services, including fuel. As such, DLA also has some advance contracts in place. In December 2018, we found that FEMA and USACE were the primary users of advance contracts.[20]

Federal Agencies Obligated at Least $5 Billion through Post-Disaster Contracts as of June 2018, but More Comprehensive Data on Disaster Contracting Obligations Would Enhance Transparency

As of June 30, 2018, federal agencies obligated at least $5 billion through post-disaster contracts to support disaster response and recovery efforts after hurricanes Harvey, Irma, and Maria and the 2017 California wildfires. USACE and FEMA awarded over three quarters of the reported obligations on post-disaster contracts. However, data on post-disaster contracting are not comprehensive due to changes in the criteria for establishing and closing a NIA code and DHS's inconsistent

[19] Public Law 84-99, § 5, codified, as amended, at 33 U.S.C. § 701n.
[20] GAO-19-93.

implementation of the criteria for closing codes. Specifically, we found DHS closed the codes for Hurricanes Harvey and Irma less than a year after the storms hit, compared to prior hurricanes when the NIA codes remained open for at least 5 years.

Federal Agencies Obligated at Least $5 Billion through Post-Disaster Contracts for the 2017 Disasters

As of June 30, 2018, federal agencies obligated at least $5 billion through post-disaster contracts in response to the three 2017 hurricanes and the California wildfires. Data on obligations for the California wildfires are limited to those contracts identified by two selected agencies in our review—FEMA and USACE—because no NIA code was established in FPDS-NG to track contracts specifically for the wildfire events at a government-wide level.[21] The obligations on post-disaster contracts accounted for more than half of the $9.5 billion in contract obligations on contracts related to the three hurricanes and the 2017 California wildfires, with the remainder of the dollars obligated on advance contracts. See Figure 6 for details on post-disaster and advance contract obligations by event.

FEMA and USACE accounted for more than three quarters of the total obligations on post-disaster contracts for the three hurricanes. Because there was no NIA code for the 2017 California wildfires, we cannot identify government-wide obligations in FPDS-NG and, therefore, do not know which agencies had the highest contract obligations for the two wildfire events. Figure 7 provides details on known obligations on post-disaster contracts, by agency.

[21] DHS makes the determination for civilian agencies on whether to open a NIA code for disaster events based, in part, on significant multi-agency federal procurement impact, according to DHS officials. They added that, given that wildfires can grow and worsen over time, the procurement impact may not be known right away. Further, they stated that no agencies requested a NIA code for the 2017 California wildfires.

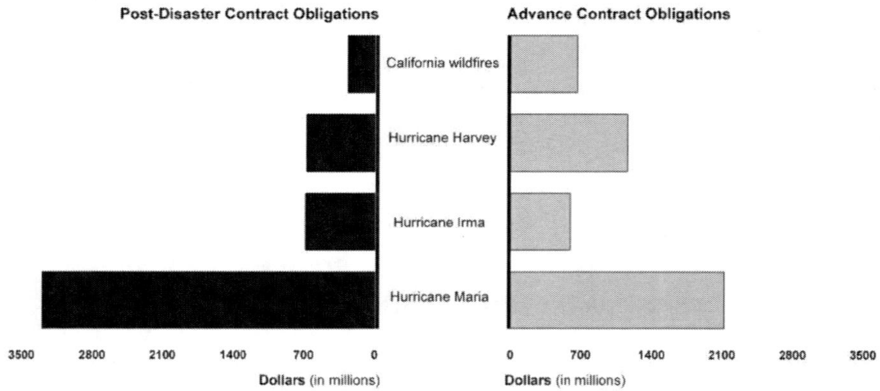

Source: GAO analysis of Federal Procurement Data System-Next Generation data. | GAO-19-281.

Notes: U.S. Army Corps of Engineers (USACE) officials told us that they did not use one of the post-disaster wildfire contracts—awarded for debris removal—and planned to deobligate the $156 million. In December 2018 USACE deobligated $140 million. This analysis includes the full $156 million that had been obligated as of June 30, 2018. Advance contract obligations are limited to Federal Emergency Management Agency (FEMA) and USACE contracts. Obligations for the 2017 California wildfires are limited to contracts identified by USACE and FEMA. FEMA officials told us that contracts awarded in support of the U.S. Virgin Islands after Hurricane Maria impacted the islands may all be coded under Hurricane Maria, regardless of whether the contract responds to needs from Hurricane Irma or Hurricane Maria.

Figure 6. Known Federal Contract Obligations in Support of the Three 2017 Hurricanes and California Wildfires, as of June 30, 2018, in Fiscal Year 2018 Dollars.

About 63 percent of the obligations on post-disaster contracts, or $3.1 billion, was for services. See Figure 8 for a breakdown of services and products by 2017 disaster.

Five services across the 2017 disasters comprised nearly 80 percent of total obligations for services on post-disaster contracts. Contracts for repair and maintenance services comprised 38 percent of total obligations on post-disaster contracts for services, largely driven by the $1 billion obligated to support the power restoration effort in Puerto Rico following Hurricane Maria. Following Hurricanes Harvey and Irma, agencies primarily awarded post-disaster contracts for management support functions, such as call center services. See Figure 9 for the top post-disaster contract services across the three hurricanes and the California wildfires.

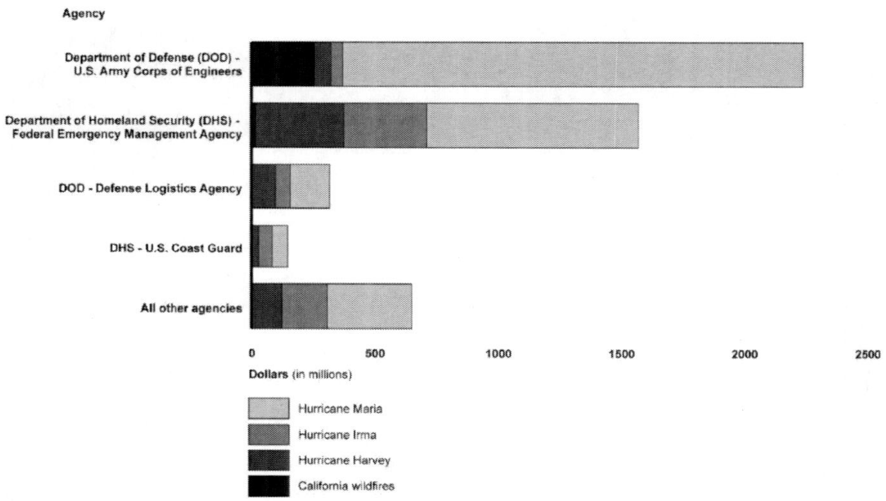

Source: GAO analysis of Federal Procurement Data System-Next Generation data. | GAO-19-281.

Note: Obligations for the 2017 California wildfires are limited to contracts identified by the U.S. Army Corps of Engineers and the Federal Emergency Management Agency.

Figure 7. Obligations on Post-Disaster Contracts in Support of the 2017 Disasters by Agency, as of June 30, 2018, in Fiscal Year 2018 Dollars.

Of the $1.8 billion agencies obligated on goods through post-disaster contracts, 28 percent was on contracts for subsistence, such as food and water. Nearly 30 percent, or more than $530 million, of all obligations on post-disaster contracts for goods was on contracts for electric wire and power distribution equipment, almost all of which was for the power mission in Puerto Rico following Hurricane Maria. See Figure 10.

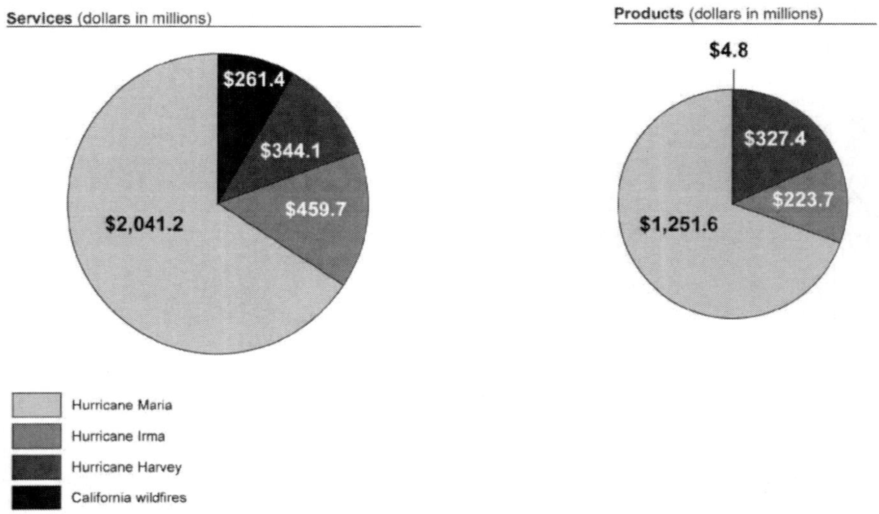

Source: GAO analysis of Federal Procurement Data System-Next Generation data. | GAO-19-281.

Note: Obligations for the 2017 California wildfires are limited to contracts identified by the U.S. Army Corps of Engineers and the Federal Emergency Management Agency.

Figure 8. Obligations on Post-Disaster Contracts for Products and Services by 2017 Disaster, as of June 30, 2018, in Fiscal Year 2018 Dollars.

Across all three hurricanes and the California wildfires, we found that the competition rate—the percentage of total obligations reported under competitive contracts—was about 75 percent for post-disaster contracts.[22] This is an increase from the past since we previously found that the competition rate in the immediate aftermath of Hurricane Katrina was about 53 percent.[23]

[22] Competitive contracts included contracts and orders coded in FPDS-NG as "full and open competition," "full and open after exclusion of sources," and "competed under simplified acquisition procedures" as well as orders coded as "subject to fair opportunity" and as "fair opportunity given," and "competitive set aside." Noncompetitive contracts included contracts and orders coded in FPDS-NG as "not competed," "not available for competition," and "not competed under simplified acquisition procedures," as well as orders coded as an exception to "subject to fair opportunity," including "urgency," "only one source," "minimum guarantee," "follow-on action following competitive initial action," "other statutory authority," and "sole source."

[23] GAO-18-335.

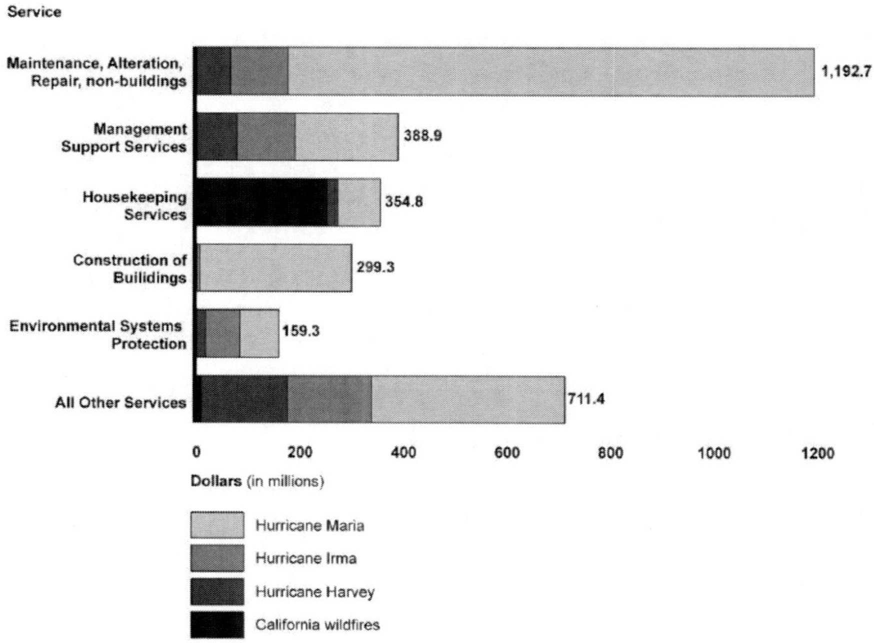

Source: GAO analysis of Federal Procurement Data System-Next Generation data. | GAO-19-281.

Note: Obligations for the 2017 California wildfires are limited to contracts identified by the U.S. Army Corps of Engineers and the Federal Emergency Management Agency.

Figure 9. Top Post-Disaster Services by 2017 Disaster, as of June 30, 2018, in Fiscal Year 2018 Dollars.

Contracting for disaster relief and recovery efforts presents unique circumstances in which to solicit, award, and administer contracts. Under the FAR, agencies are generally required to use full and open competition when soliciting offers, with some exceptions. As discussed earlier, an agency may award a contract without full and open competition, for example when the need for goods and services is of such an unusual and compelling urgency that the federal government faces the risk of serious financial or other type of loss, unless the agency is permitted to limit the number of sources from which it solicits offers ("urgency exception").[24]

[24] FAR § 6.302-2.

When using the urgency exception, the FAR requires agencies to request offers from as many potential sources as practicable.[25]

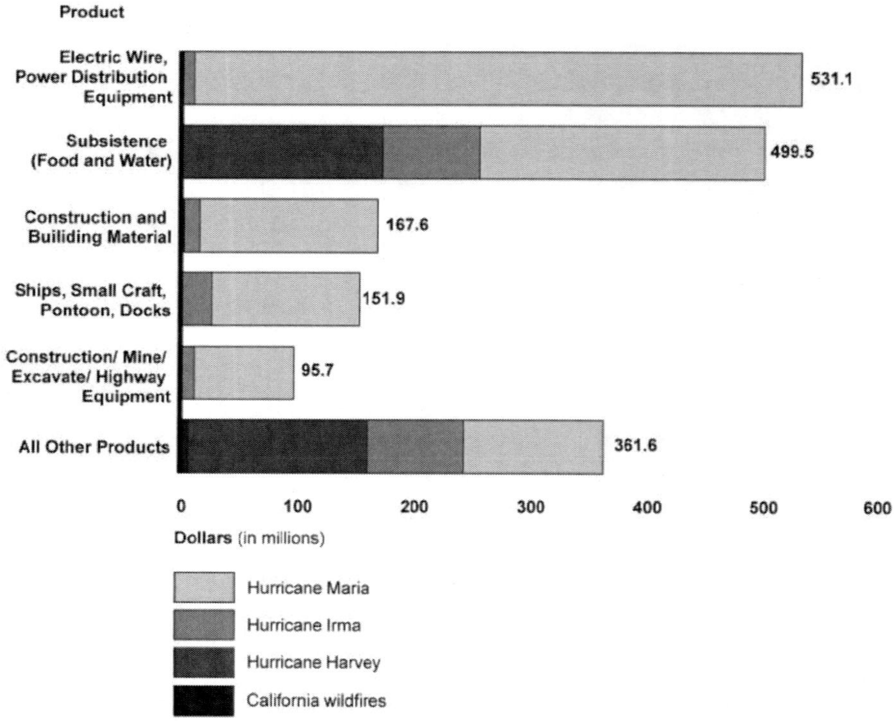

Source: GAO analysis of Federal Procurement Data System-Next Generation data. | GAO-19-281.

Note: Obligations for the 2017 California wildfires are limited to contracts identified by the U.S. Army Corps of Engineers and the Federal Emergency Management Agency.

Figure 10. Top Post-Disaster Products by 2017 Disaster, as of June 30, 2018, in Fiscal Year 2018 Dollars.

Based on FPDS-NG data, we found that about 47 percent of obligations on post-disaster contracts were on contracts citing the urgency exception, with 63 percent of those obligations on contracts coded in FPDS-NG as using "limited competition." Among our selected contracts,

[25] Pub. L. No. 109-295, § 695(b) (2006), as implemented in HSAR §§ 3006.302-270, 3006.303-270; FAR §§ 6.302-2, 6.303-1.

we also found that contracting officers implemented the urgency exception to seek offers from as many sources as possible in different ways. Of the 11 contracts in our sample that cited the urgency exception, five included abbreviated award time frames in the justification documentation.

More Comprehensive Data Could Provide Increased Transparency on Disaster Contracting

The full extent of disaster contracting—for both advance and post-disaster contracts—related to the 2017 disasters is unknown due to changes in the criteria for establishing and closing a NIA code in FPDSNG and DHS's inconsistent implementation of the updated criteria for closing codes. The NIA code data element in FPDS-NG was established following landfall of several major hurricanes in 2005 to enable consistent tracking of emergency or contingency-related contracting. Contracting officers select the applicable NIA code in FPDS-NG when entering related contract information into the system. Officials at GSA—the agency responsible for operating and maintaining FPDS-NG—stated there is little to no cost or administrative burden associated with establishing or maintaining a NIA code.[26]

Based on a memorandum of agreement (the agreement), GSA, DHS, and DOD are jointly responsible for determining when a NIA code should be established and closed. DHS delegated its role, on behalf of civilian agencies for disaster or emergency events, to its Office of the Chief Procurement Officer (DHS OCPO), and DOD, on behalf of military departments and defense agencies for contingency operations, delegated its

---

[26] Defense Pricing and Contracting Office and DHS OCPO officials expressed anecdotal concerns about the reliability of NIA code data when multiple codes are open at the same time, stating that more codes introduce more potential for error. Based on our review of the NIA code data, we did not find significant reliability concerns related to miscoding events. Further, FPDS-NG includes data elements with thousands of codes available for selection—such as the North American Industry Classification System and Product Service Code fields. Based on results of agency sampling reported in the Fiscal Year 2017 Federal Government Procurement Data Quality Summary, these fields are generally coded accurately.

role to the Defense Contract and Pricing office. The agreement outlines criteria DHS and DOD should consider in making determinations to establish and close a NIA code.[27] We identified changes in the criteria for establishing and closing a NIA code between a June 2012 agreement and a June 2018 update that superseded and replaced it.

Table 2. Changes in the Criteria for Establishing and Closing National Interest Action (NIA) Codes between the 2012 and 2018 Memorandums of Agreements

Criteria in the 2012 memorandum of agreement	Change from 2012 to 2018		Examples of potential effect of the change for emergency or disaster events GAO identified
Establishing a NIA code	Only "high visibility" events with certain characteristics will be considered for a NIA code in the Federal Procurement Data System-Next Generation (FPDS-NG)	The updated criteria specifically state that the federal mobilization should be nationwide, not just regional.	DHS may not request NIA codes when the event impacts one region within a state or only one state, such as wildfires in California or hurricanes in Hawaii.
		The updated criteria added that DHS should consider the significant multi-agency procurement impact related to an event.	DHS may not request a NIA code for an event unless multiple agencies are awarding contracts in response to it.
	NIA code is opened when emergency acquisition flexibilities are authorized for an agency	The updated criteria added that a new NIA code may only be requested when more than one agency exercises, or intends to exercise, the special emergency procurement authority to increase procurement thresholds, such as simplified acquisition and micro-purchase thresholds, for an event.[a]	DHS may not request a NIA code for an event unless at least two agencies have exercised the special emergency procurement authority to increase certain procurement thresholds

[27] According to the agreement, DHS is primarily responsible for requesting the establishment and closing of NIA codes for civilian agency emergency or disaster events, while DOD does so for defense contingency operations.

Criteria in the 2012 memorandum of agreement	**Change from 2012 to 2018**		**Examples of potential effect of the change for emergency or disaster events GAO identified**
Closing a NIA Code	Code is closed when national interest and national mobilization has ended	The updated agreement removed the criteria that the NIA code is closed when the event no longer has high visibility and other interest in the information. The updated criteria added that a NIA code can be closed when agencies have permitted procurement thresholds to return to their pre-disaster levels; and that remaining or predicted contracting activity applicable to the NIA has become routine and, using judgement, that it is no longer prudent to track contract actions using a system-wide NIA value.	DHS can allow a NIA code to expire regardless of the high visibility of the NIA code, even when multiple agencies continue to routinely award contracts related to the event, such as for Hurricanes Harvey and Irma.
	Apply sound judgment and management expertise related to remaining contracting activity to determine when to end date the NIA code	The updated criteria adds that the agency can close the NIA code when the number of contract actions is so small that contracting activities have adequate alternate methods other than a system-wide NIA code to identify disaster procurements.	The addition of the criteria relies on agencies to have the capacity to track contract actions by disaster or contingency event by some means other than the NIA code. Further, the change to alternate methods would not provide for system-wide, publicly available data and transparency related to the event.

Source: GAO analysis of the 2012 and 2018 NIA code Memorandums of Agreements | GAO-19-281.

[a]Contracts awarded under the simplified acquisition threshold—generally $150,000 during our review—may use simplified acquisition procedures. Purchases under the micro-purchase threshold— generally $3,500 during our review—do not need to meet certain requirements for procurements above this threshold and contracting officers are not required to input these into FPDS-NG. Agencies may increase these thresholds for procurements to be used in support of an emergency or major disasters.

According to DHS OCPO officials, the agencies updated the agreement to incorporate lessons learned (such as adding that events should have a procurement impact as criteria for establishing a NIA code), and because it had not been revisited in 6 years. See Table 2 for criteria from the agreements, changes in 2018, and examples of potential implications of those changes that we identified related to emergency or disaster events. The June 8, 2012 agreement criteria applied to the establishment of NIA codes for the 2017 disasters, while the June 1, 2018 updates applied to determinations to close or extend the NIA codes after this date for the 2017 disasters. DHS OCPO requested that a NIA code be established for each of the 2017 major hurricanes (Harvey, Irma, and Maria). However, the codes for Harvey and Irma closed on June 30, 2018, less than a full year after the hurricanes hit. The code for Maria was scheduled to close on December 15, 2018, and in August 2018 we began raising questions about the planned or actual NIA code closures for the three 2017 hurricanes. Since December 2018, DHS OCPO provided two additional extensions for Maria, with the code now valid through June 15, 2019, about 21 months after that hurricane made landfall. In contrast, the NIA code for Hurricane Sandy, which made landfall in October 2012, remained open until December 2017, more than 5 years after the disaster. The NIA code for Hurricane Katrina, which made landfall in August 2005, remained open until August 2018, 13 years after the disaster. We observed that DHS OCPO requested NIA codes for Hurricanes Florence and Michael in 2018, although we did not review the data associated with those events. After we sent this chapter to the agencies for comment on February 15, 2019, the agencies allowed the codes for Florence and Michael to expire, on March 15, 2019 and April 12, 2019, respectively.

DHS OCPO officials offered several different rationales to support their decision to close the NIA codes for the 2017 hurricanes and cited the changes to the criteria in the 2018 agreement for closing the codes.

However, we found that these rationales were inconsistent with the criteria in the agreement, did not consider key user needs, and did not fully explain the decisions to close these codes. For example:

- DHS OCPO officials told us that NIA codes for disasters should be closed when agencies no longer use the special emergency procurement authority such that the procurement thresholds—such as the simplified acquisition and micro purchase thresholds—return to the general (non-emergency) procurement thresholds in the FAR.[28] Further, when FEMA requested to keep the codes open, DHS OCPO questioned why agencies would need to continue tracking with a NIA code after the thresholds had returned to general procurement thresholds. DHS officials stated that the updated agreement put an emphasis on this criterion; however, our analysis indicated that was not consistent with 2018 agreement, which includes multiple criteria and is not limited to this factor. Further, the agreement does not provide additional emphasis on one criterion over others.

- DHS OCPO officials stated that the purpose of the NIA code is to track federal procurement related to response, not recovery efforts. However, both the 2012 and 2018 agreements specifically state that the NIA code is intended to track disaster response and recovery efforts. Further, according to the National Response Framework and National Disaster Recovery Framework, we found that there are no clear lines of distinction between the start and end date of these two efforts, and often these stages of the process overlap. Additionally, FEMA officials from the Recovery Support Function Leadership Group's Program Management Office stated that they use the NIA code to track government-wide contracting related to recovery efforts. The Recovery Support Function Leadership Group, an interagency body chaired by FEMA, tasked

---

[28] Agencies can exercise the use of special emergency procurement authorities within 41 U.S.C § 1903(a)(b), (and as implemented by DHS FAR Class Deviation 17-02 and DOD FAR Class Deviation 2017-O0007) to increase the micro-purchase threshold and simplified acquisition thresholds for procurements in support of domestic emergency or disaster relief efforts. Generally, the simplified acquisition threshold was $150,000 and the micro-purchase threshold was $3,500 under the FAR. The National Defense Authorization Act for Fiscal Year 2018 raised the simplified acquisition threshold to $250,000 and the micro-purchase threshold to $10,000. Pub. L. No. 115-91, §§ 805, 806; for the increase to the micro-purchase threshold for DOD see John S. McCain National Defense Authorization Act for Fiscal Year 2019, Pub. L. No. 115-232 § 821. FAR Case 2018-004, Increased Micro-Purchase and Simplified Acquisition Thresholds (open as of Jan. 25, 2019). On May 30, 2018, DHS issued a FAR class deviation implementing these changes while the FAR is updated.

the Program Management Office with providing accountability and transparency of projects and outcomes for the 2017 disasters, among other things.

- DHS OCPO officials pointed to the Digital Accountability and Transparency Act of 2014 as providing alternatives to FPDS-NG. The Digital Accountability and Transparency Act of 2014 required improvements in the quality of data on federal spending, including disaster spending, by making data more accessible and transparent, such as by improving the quality of data submitted by federal agencies to USA Spending—an online tool that tracks federal grant, loan, contract, and other awards. However, we found that USA Spending provided some information on contract obligations using disaster response and recovery funds but does not separate obligations by disaster event. Further, our prior work on the Digital Accountability and Transparency Act of 2014 has found limitations with the data agencies provide, notably the completeness and accuracy of data.[29] Specifically, we found that agencies routinely provided award descriptions in an abbreviated way and lacked clarity needed to compare data across the federal government. Moreover, we found inconsistencies in agencies' ability to track contract actions by disaster. While FEMA has the capacity to provide contract information by disaster through a centralized contract tracking tool, USACE officials stated that they use a decentralized tracking process where they reach out to the districts and centers to identify and track disaster contracts without a NIA code.
- Prior to the June 30, 2018 decision to close the NIA codes for Harvey and Irma, DHS OCPO officials told us they found that the number of actions FEMA was making for these events had decreased. Our analysis of the NIA codes showed that components

---

[29] GAO *DATA Act: Reported Quality of Agencies' Spending Data Reviewed by OIGs Varied Because of Government-wide and Agency Issues*,GAO-18-546 (Washington, D.C.: July 23, 2018), and; *DATA Act: OMB, Treasury, and Agencies Need to Improve Completeness and Accuracy of Spending Data and Disclose Limitations*, GAO-18-138 (Washington, D.C.: Nov. 8, 2017).

across ten departments, including within DHS and DOD, were executing contracts related to Harvey and Irma in June 2018. When we requested supporting documentation and analysis, DHS OCPO officials provided some correspondence with FEMA but did not provide government-wide data analysis to identify what other agencies were awarding and executing contracts related to these events. DHS OCPO officials stated they also sought input from DOD through the Defense Pricing and Contracting Office on whether to keep the codes open. According to DHS officials, DOD deferred to DHS on the decision because DHS was responsible for establishing the codes. Further, DOD officials did not provide evidence that would allow us to determine whether they assessed which defense components were executing contracts related to these events or sought the input of the components that were doing so, such as USACE and the Navy.

FPDS-NG—a public, government-wide database of federal procurements—offers a resource the federal government can use to create recurring and special reports for key users, such as the President, Congress, executive agencies, and the general public. The NIA code in FPDS-NG provides consistent tracking and government-wide visibility into contracting related to disaster events through a publicly available database. Without clear criteria for establishing and closing NIA codes that consider the needs of data providers and users, such as FEMA, and the high visibility of the event being tracked and a mechanism to ensure consistent implementation of these criteria, insight into disaster contracting may be limited. Additionally, federal internal control standards state that management should use quality information, communicate quality information internally, and communicate quality information externally to achieve objectives.[30] Management should accomplish this by considering appropriate methods for communicating externally, such as to the President, Congress, and the general public.

---

[30] GAO, *Standards for Internal Control in the Federal Government*, GAO-14-704G (Washington, D.C.: September 2014).

As noted above, the 2018 agreement no longer includes the 2012 criteria that a NIA code can be closed if the NIA no longer has high visibility and there is no other interest in the NIA code. In our discussions with officials, DHS OCPO could not provide a rationale for these changes and the rationale is also not included in the updated agreement. Prior to DHS OCPO's decision to close the codes for Hurricanes Harvey and Irma, a senior FEMA procurement official requested that they remain open, in part because of the high visibility of these events. As such, this official stated that there will be continued interest in the 2017 hurricanes including inquiries from Congress, which will require agency officials to pull data for interested parties, as that data can no longer be tracked and identified through public databases, such as FPDS-NG and USASpending. DHS OCPO officials denied FEMA's request, pointing to the criteria in the 2018 agreement, which does not include consideration of the visibility of the event or key user needs. As the federal agency responsible for coordinating disaster response and recovery, FEMA is well positioned to understand the level of national and political interest in tracking procurement information for a disaster or emergency event. Yet, it is unclear why neither the 2012 nor the updated 2018 agreements included a role for or consideration of key users, such as FEMA and Congress. Further, as noted above, FEMA program officials expressed concern over closing the Harvey and Irma codes because they had planned to use the codes to assess recovery efforts for the 2017 disasters.

As we have previously reported, it can take years to fully account for federal contract obligations related to response and recovery after a hurricane.[31] Once a NIA code is closed, there is no publicly available, government-wide system to track contract obligations for specific events. Moreover, DHS OCPO officials were unable to provide data analysis conducted using available data from prior events to determine historical patterns in federal contracting obligations for disasters prior to closing the codes for Hurricanes Harvey and Irma. Figure 11 illustrates the lack of insight we have into disaster contracting activities related to the 2017

[31] GAO-18-335.

hurricanes, in comparison to what we know about prior storms with high federal procurement obligations.

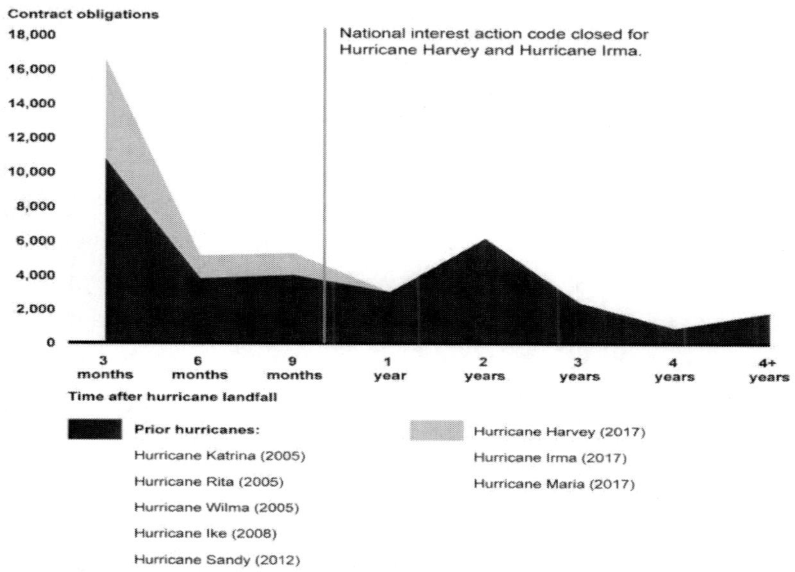

Source: GAO analysis of Federal Procurement Data System-Next Generation data. | GAO-19-281.

Note: This figure includes reported obligations from August 2005 through September 2018. The national interest action codes for Hurricanes Harvey and Irma were closed about 10 months after these storms made landfall. As such, obligations for these events made after June 30, 2018 were identified using the description field in FPDS-NG, and do not provide the full extent of procurement activities for these events.

Figure 11. Contract Obligations over Time for Prior Hurricanes and the 2017 Hurricanes, in Fiscal Year 2018 Dollars (dollars in millions).

Further, using the description field in FPDS-NG, we found that between July 1 and September 30, 2018, after the NIA codes were closed, agencies obligated at least $136 million on contracts for Hurricane Harvey and $123 million on contracts for Hurricane Irma. While this provides some important insights regarding the continued contracting activity related to these hurricanes, the description field in FPDS-NG cannot be relied on to provide a full picture. Some agencies may include event-specific information in the description field; however, we found that, for the 2017 hurricanes, about 65 percent of contract obligations linked to a NIA code did not include event-specific information in the description. Without reopening the NIA codes for Hurricanes Harvey and Irma, and, to the extent practicable, retroactively populating the NIA codes for contract actions supporting response and recovery for these hurricanes during the period they were closed, decision makers are missing important information to understand the procurement impact of these disasters. Retroactively entering NIA code information is not unprecedented. For example, based on our analysis, the NIA codes for the 2005 hurricanes were established in October 2005, and contracting officers retroactively entered data for contracts related to these events which occurred as early as August of that year to enable full insight into contracting for these disasters.

Challenges in Planning Post-Disaster Contracts Hindered Response and Recovery Efforts

Based on the contracts we reviewed and officials we spoke with responsible for the planning of these contracts, we found that agencies experienced challenges planning for post-disaster contracts, especially when it came to contracting with local vendors. Additionally, FEMA also experienced challenges with requirements development—in that program officials did not always provide well-defined or sufficiently specific requirements for post-disaster contracts. However, FEMA has taken steps to address its challenges with requirements development, but it is too soon to tell the extent to which these steps will address the challenges we identified.

Agencies We Reviewed Experienced Challenges Contracting with Local Vendors

The Stafford Act, as amended, contains mechanisms to provide federal assistance to affected communities in the aftermath of a major disaster and jump-start the economy through the award of contracts to local businesses in the disaster area. The FAR, which implements the Stafford Act requirement to provide a preference in awarding disaster response contracts to local firms where feasible and practicable, offers contracting officers some flexibility to make local awards (see sidebar). We found that for the contracts we reviewed, agencies did not consistently meet requirements related to contracting with local vendors or were confused about how to implement certain FAR requirements. Specifically, agencies did not always:

Steps to Implement Local Vendor Preference, as Outlined in the Post-Katrina Emergency Management Reform Act and the Federal Acquisition Regulation (FAR)

Step 1: Identify the set-aside area in accordance with FAR § 26.202-1—Local Area Set-Aside and § 6.208—Set-asides for Local Firms During a Major Disaster or Emergency

Step 2: Conduct market research to determine whether there are qualified vendors in the set-aside area.

Step 3: Issue a solicitation that provides for local vendor preference to the extent feasible and practicable either through the use of a set-aside or an evaluation preference.

Step 4: Review offers based on evaluation criteria in the solicitation. If using a local area set-aside, review information from potential vendors to determine if they reside or primarily do business in the set-aside area in accordance with FAR § 52.226-3—Disaster or Emergency Area Representation.

Step 5: Award contract to qualified vendor. If the vendor selected is not local or no qualified vendors are in the set-aside area, justify the decision in writing.

Source: GAO Analysis of the Post-Katrina Emergency Management Reform Act and Federal Acquisition Regulation | GAO-19-281.

- accurately define the geographic area for a local area set-aside,
- determine that a vendor resides or primarily does business in the local area, or
- justify in writing awards that they made to vendors outside the set-aside area.

Some Officials We Interviewed Were Not Consistently Aware of the Regulatory Definition of Local Area

For the contracts we reviewed, contracting officials at FEMA correctly identified the local area for six set-aside contracts across the three hurricanes, and USACE correctly identified the local area for two set-aside contracts in Puerto Rico.[32] However, based on the interviews we conducted during our review, USACE contracting officials were not consistently aware of the specific regulation for doing so and did not correctly identify the local area for two other USACE contracts awarded in support of the California wildfires. When awarding a local area set-aside or using an evaluation preference for local vendors, FAR § 26.202-1 states that a major disaster area can span several counties in several contiguous states, but need not include all the counties in the disaster area, and cannot extend beyond the counties designated in a Presidential disaster declaration.

Figure 12 provides an example of a disaster declaration that depicts which counties could be included in the set-aside area.

For all six local area set-aside FEMA contracts—awarded in response to Hurricanes Harvey, Irma, and Maria—we reviewed, FEMA officials defined the local area in accordance with regulation. This was an improvement from what we previously found. Specifically, in 2015, we found that FEMA contracting officers were confused about the definition of the set-aside area and recommended that the FEMA Administrator provide new or updated guidance to ensure all contracting officers are aware of requirements concerning contracting with local vendors, among other things.[33] DHS concurred, and FEMA updated its annual disaster

---

[32] Among our selected contracts, only FEMA and USACE used local area set-asides.
[33] GAO-15-783.

contracting webinar training to reiterate the requirement and clarify how to determine the geographic area using the disaster declaration.

California Wildfire Disaster Declaration as of 11/28/17

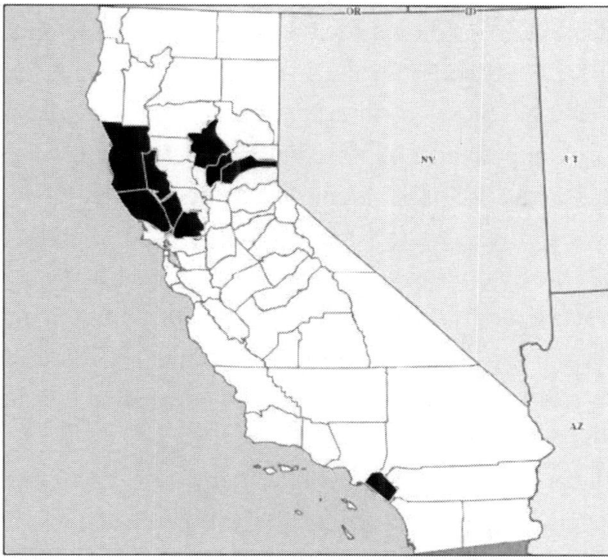

■ Counties named in the disaster declaration for the California wildfires

Source: Federal Emergency Management Agency (FEMA). | GAO-19-281.

Figure 12. Presidential Major Disaster Declaration Map Depicting the Local Set-Aside Area, 2017 California Wildfires.

For the two local area set-aside USACE contracts awarded, officials responsible for those contracts told us that when awarding these contracts, they were not aware of the regulatory requirements for defining the geographic area of the local area set-aside. However, as the presidential disaster declaration for Hurricane Maria included the entire island of Puerto Rico, the local set-aside area covered the entire island. As a result, officials met the set-aside area requirement in accordance with regulation, even though they noted that they were not familiar with the requirement at the time. Officials told us they became aware of the regulation after conducting research pursuant to a protest related to the use of local vendor preference.

We also reviewed two other USACE contracts that were used to support the debris removal mission following the California wildfires. Contracting officials stated that they conducted market research on the availability of local contractors, and they ultimately did not find qualified local firms. However, based on a review of contract file documentation, we found that USACE officials did not identify the local area in accordance with regulation for these contracts. Instead they used congressional districts that overlapped with impacted areas to identify the local area. We found that the areas USACE identified included areas outside of the geographic area defined by the presidential disaster declaration for the California wildfires. Contracting officials responsible for these debris removal contracts stated they were not aware of a policy or regulation for how to identify the geographic area for a local area set-aside, but that their office had internally determined the use of congressional districts impacted by a disaster to be the preferred method.

A senior USACE official told us that there is no agency supplemental guidance or related training regarding the use of local vendor preference for contracts supporting disaster recovery and response, only that they expect USACE contracting officials to comply with the FAR. Without additional guidance or related training, contracting officers may be unaware of how to define the geographic area for a local area-set aside in accordance with regulation and may miss opportunities to support improving the local economies of disaster impacted areas by giving preference in awarding contracts to local vendors to the extent feasible and practicable, per the Stafford Act.

Regulation for Determining Whether a Vendor Resides or Primarily Does Business in the Set-Aside Area Presents Challenges

Despite contracting officers having a high degree of discretion to determine that an offeror qualifies as a "local firm,"—that is, a firm that resides or primarily does business in the designated set-aside area—contracting and legal officials at both FEMA and USACE told us they were unsure what or how much information is sufficient to determine that an offeror qualifies as a local firm under the FAR. After contracting officials have identified the geographic boundaries of the local "major disaster or emergency area" and included required clauses in the solicitation and issued it as a local area set-aside, offerors must represent in their offer that they reside or primarily do business in the set-aside area. Specifically, FAR § 52.226-3(c) outlines two criteria a contracting officer should use to determine whether an offeror is to be considered "local." If an offeror does not meet these first two criteria, FAR § 52.226- 3(d) provides eight additional criteria contracting officers may consider to make this determination (see sidebar).

Determining that an Offeror Qualifies as Local under FAR § 52.226-3

Determine if the vendor meets the criteria under FAR § 52.226-3(c)

1) An offeror is considered to reside or primarily do business in the set-aside area if, during the last 12 months, 1) the offeror had its main operating office in the area; and 2) that office generated at least half of the offeror's gross revenues and employed at least half of the offeror's permanent employees.
2) If the offeror does not meet the criteria under FAR § 52.226-3(c) consider other factors listed in FAR § 52.226-3(d) including:
    1) Physical location(s) of the offeror's permanent office(s) and date any office in the set-aside area(s) was established;
    2) Current state licenses;
    3) Record of past work in the set-aside area(s);
    4) Contractual history the offeror has had with subcontractors and/or suppliers in the set-aside area;
    5) Percentage of the offeror's gross revenues attributable to work performed in the set-aside area;
    6) Number of permanent employees the offeror employs in the set-aside area;
    7) Membership in local and state organizations in the set-aside area; and
    8) Other evidence that establishes the offeror resides or primarily does business in the set-aside area.

Source: GAO analysis of Federal Acquisition Regulation | GAO-19-281.

Of the eight local area set-aside contracts we reviewed, two were impacted by bid protests—which is when an offeror challenges an award or proposed award of a contract or a solicitation—related to the FAR criteria for determining that an offeror qualifies as a local firm. The following protests show examples of the criteria agencies reviewed to determine whether a firm resided or primarily did business in a set-aside area.

- FEMA contract for food: In a protest of the award of a contract for food on the basis that FEMA improperly determined the protester failed to meet the requirements in FAR§ 52.226-3(d), the protester stated it met the requirements of FAR § 52.226-3(d), because it had (1) done past work in the set-aside area; (2) maintained a warehouse in the set-aside area; (3) maintained a contractual history with subcontractors in the set-aside area; and (4) maintained a current state license and filed a franchise tax return. FEMA denied, the protest stating that the evidence the protester provided was not sufficient to qualify as "residing or primarily doing business" in the local area.
- USACE Blue Roof contract: To support the Blue Roof mission—which provides temporary blue plastic roofs for disaster-impacted residences to prevent further damage and allow homeowners to arrange for permanent repairs—following Hurricane Maria in Puerto Rico, contracting officials awarded two post-disaster contracts. In a protest of the awards filed with GAO, the protestor argued, among other things, that one of the awardees did not meet local firm criteria in FAR § 52.226-3(c).[34] USACE had assessed information on the awardee, including its local business address in the System of Award Management and other documentation of prior work in Puerto Rico, prior to award and determined that the awardee met Stafford Act criteria for award to a local vendor. USACE officials told us that, after the protest was filed, they further assessed information on the awardee in question and determined that it was a subsidiary of a larger national company. According to USACE officials, in order to quickly continue work

[34] A bid protest is a challenge to the terms of a solicitation or the award of a federal contract. GAO's Procurement Law Division adjudicates bid protests against procurement actions by a federal agency.

on the Blue Roof mission, which had increased in scale, USACE negotiated pricing with the protestor while the protest was ongoing and made a third award under the solicitation. The protestor withdrew the protest.

Contracting and legal officials at FEMA and USACE described difficulty in determining whether a vendor resides or primarily does business in the local set-aside area and cited a lack of clarity and different interpretations of the FAR. Based on conversations with the agencies' legal officials, we found that USACE and FEMA applied the eight criteria in FAR § 52.226- 3(d) differently. FEMA officials told us that in determining whether a firm is local, if the first two criteria are not met, they evaluate an offeror's information related to the eight criteria in FAR §52.226-3(d) to see if the first two criteria can be met with this additional information. They added that they look to see if the firm's main operating office is in the set-aside area and if that office generated at least half of the offeror's gross revenues and employed at least half of its permanent employees, but stated that the eight criteria do not need to be met within the last 12 months. Alternatively, USACE officials told us that in determining if a firm is local, if the first two criteria are not met, they evaluate an offeror's information against the eight criteria in FAR § 52.226-3(d) independent of the two criteria described under FAR § 52.226-3(c). Legal officials at both USACE and FEMA stated that the FAR criteria should be clarified. Further, agencies' varying application of the criteria increases the risk that an offeror may be considered local by some agencies, but not others.

FEMA legal officials told us that contracting officers have been instructed to ask offerors for information on a local firm status in post-disaster solicitations. USACE legal officials explained that it is not always clear what specific information or documents provide the necessary information to meet the criteria under FAR § 52.226-3. For example, it may not be clear what documentation adequately demonstrates the number of permanent employees the offeror employs in the set-aside area, or the percentage of the offeror's gross revenue earned in the set-aside area. The Office of Federal Procurement Policy provides overall direction of government-wide procurement policies, regulations, procedures, and forms for executive agencies. However, Office of Federal Procurement Policy staff told us that they have not provided additional guidance or clarification related to this FAR clause.[35] Federal internal control standards state that management should use quality information to achieve objectives. Management should accomplish this by identifying information requirements, collecting relevant data from reliable sources, and processing data into quality information to be communicated internally and externally.[36] Without clarifying guidance, contracting and legal officials will likely continue to have varying interpretations on how to implement the FAR criteria for determining that an offeror qualifies as a local firm.

Some Agencies We Reviewed Did Not Consistently Write Justifications for the Use of Non-Local Vendors

When contracts for major disaster or emergency assistance activities are not awarded to local vendors, the Stafford Act, as implemented in the FAR, requires that the decision be justified in writing in the contract file. Contracting officers at three of the four agencies included in our review—FEMA, USACE, and the Coast Guard—did not consistently justify in writing the award of selected contracts to non-local vendors. Specifically, 12 of the 14 contracts in our review that were not awarded to local vendors did not contain the required written justifications in the files (see Table 3).

DLA included written justifications for the use of non-local vendors, as required. After the 2017 disasters, FEMA identified the absence of justifications for the use of non-local vendors as an area for improvement. According to FEMA officials, they subsequently released guidance and a pre-solicitation memorandum to assist contracting officers in identifying what documentation related to local vendor preference is required in a

[35] In 2007, the FAR Council issued an interim rule that proposed amending the FAR to include the current criteria under § 52.226-3. In doing so, the council wrote about some issues pertaining to FAR § 52.226-3, for example, the exclusion of corporate branch offices for the purposes of local area set-aside contracts, citing the intent of Congress to favor firms based in the local area who hire local people. At that time, the council invited public comment, but received none, and the interim rule was adopted as a final rule in 2008 without changes to the criteria.

[36] GAO-14-704G.

contract file. FEMA officials told us they expect these steps will improve compliance with the requirement to document the justification for using non-local vendors going forward.

Table 3. Overview of Selected Contracts Use of Local Vendor Preference

Agency	Total Selected Contracts	Contracts Awarded to Local Vendors	Contracts Awarded to Non-local Vendors	Contract Files that Included Documentation of the Use of a Non-local Vendor
Federal Emergency Management Agency	12	5	7	0
U.S. Army Corps of Engineers	7	3	4	0
Coast Guard	2	1	1	0
Defense Logistics Agency	2	0	2	2
Total:	23	9	14	2

Source: GAO analysis of contract files. | GAO-19-281.

While the Coast Guard provided a memorandum ahead of the 2017 disaster response that addressed the use of local vendors, it did not reference the requirement under the Stafford Act, as implemented in the FAR, to justify in writing the use of non-local vendors. A senior USACE official told us the agency had not issued any guidance to address requirements for contracting with local vendors and was not aware of any guidance issued at the department level. USACE legal officials noted the lack of written justification may be due to abbreviated timeframes under which post-disaster contracts are awarded. However, we found that USACE contracts included consolidated justification documents outlining rationales for the use of limited competition or abbreviated solicitation timeframes, but they did not include justifications for the use of non-local vendors. Without additional guidance or tools, contracting officials may not be aware that they are required to include written justifications for the use of non-local vendors in contract files, and federal agencies are at risk of not complying with the Stafford Act requirement to do so.

FEMA Has Begun to Address Challenge with Requirements Development for Post-Disaster Contracts

Contracting officers responsible for the FEMA contracts we selected and senior procurement officials stated that during disaster response they received post-disaster requirements packages that were lacking in technical specificity or were otherwise deficient, but FEMA has begun to address this challenge. Program officials communicate contract requirements to contracting officers through requirements documents that include, among other items, a statement of work describing goods or services to be provided by an offeror, market research, and an independent government cost estimate. Contracting officials explained that when they received deficient documents, they had to conduct additional work to refine the requirements before soliciting for the contract—such as spending time assisting program officials to develop the required documentation. This additional work may add time to already tight award time frames for post-disaster contracts. When compared to large dollar value acquisitions, post-disaster contracts are awarded on significantly abbreviated time frames. For example, among the 12 FEMA contracts we assessed, time frames between the submission of a resource request and award date ranged from 1-26 days.[37] This is faster than suggested; FEMA's Procurement Administrative Lead Time guidance suggests preparation time frames of 60-300 days for new procurements based on the nature and value of an action.

We found instances where FEMA program offices provided inaccurate or untimely estimates of the quantities of goods or services needed for the contracts we reviewed, in some cases leading to additional time and efforts spent to meet the need. For example:

- After Hurricane Harvey, FEMA awarded contracts to supply a food bank. Officials told us the initial requirement from the food bank through the program office to the contracting officer was

[37] This analysis included the 12 FEMA contracts out of our sample of 23 contracts.

expressed in terms of "truck loads" but did not specify, for example, how large the truck should be, or how many pallets should be loaded per truck. FEMA ultimately awarded three contracts to meet the post-disaster need—the first contract had a period of performance of 4 days and, according to FEMA officials, was intended to meet initial needs for food while the program and contracting officials determined the full scope of the requirement. The second contract—a $37 million contract with a period of performance of 52 days—was intended to fulfill the remaining requirement. However, due to miscommunication of the requirement as documented in the contract files and according to a program official responsible for the contracts, FEMA needed to award a third contract for an additional 2.5 months and $23 million to meet the need. Due to the value of the contracts, FEMA deemed that the subsequent contract required a new solicitation and award, rather than a modification to the existing contracts, thereby increasing the time and effort required of procurement personnel to meet the post-disaster need for food.

- In response to Hurricane Maria, FEMA awarded four post-disaster contracts for self-help tarps—which are used to cover small areas of roof damage. Of these contracts, two were terminated for convenience, both of which were included in our sample. The terminations were due in part to a national supply shortage. FEMA officials told us that under one of the contracts included in our review, at the request of the Commonwealth of Puerto Rico through program officials, FEMA ordered 500,000 40-foot-by-40-foot tarps, which differ from the size of the tarps normally ordered and stocked by the agency. Due to the supply shortage, FEMA received none, but officials noted that the impact of not receiving the tarps was minimal because the agency had initially overestimated the total number of tarps needed.

Since the 2017 disasters, FEMA has started to address the issues with requirements development. Specifically, in 2018, FEMA officials told us the agency used portfolio managers in the field to assist with developing requirements for disaster response. Previously, in 2017, portfolio managers told us they supported the National Response Coordination Center but did not deploy to the disasters. Organizationally housed within FEMA's OCPO, portfolio managers we spoke with told us they provide general templates for and guidance on acquisition documents for program officials to use and are primarily responsible for supporting steady-state acquisitions included in FEMA's Master Acquisition Planning Schedule. Additionally, portfolio managers told us they provide informal, optional, "brown bag" training sessions for program officials. FEMA OCPO officials told us that they receive more requests for portfolio manager assistance than they can support, as the portfolio management section only maintains up to six staff. FEMA OCPO officials noted, however, that the agency expected to award an acquisition support contract to expand portfolio management capabilities. While the use of portfolio managers is an important step, it is too soon to tell the extent to which the use of portfolio managers in the field will address FEMA's challenges with requirements development for post-disaster contracts.

Agencies in Our Review Have Identified Some Lessons Learned in Disaster Response, but Interagency Contracting Coordination and FEMA Workforce Challenges Remain

The agencies we reviewed each have a process for identifying lessons learned following a disaster, and we found they used these processes for the 2017 disasters. While agencies have identified actions they plan to take in response to the lessons they found following the 2017 disasters, additional challenges remain. Specifically, the agencies in our review encountered interagency contracting coordination challenges during the mission assignment process. Further, FEMA identified disaster contracting workforce shortages.

Selected Agencies Have Processes for Identifying Lessons Learned

FEMA, USACE, Coast Guard, and DLA each have processes for identifying lessons learned within their agencies through after-action reports. These reports identify lessons learned and areas for improvement and may be completed following a training exercise or a real-world event. Through these processes, agencies identified lessons learned during the 2017 disasters.

Table 4. Selected Agencies' Practice or Requirement for Identifying Lessons Learned from the 2017 Disasters, Key Contracting and Mission Assignment Findings, and Proposed Actions[a]

Agency	Agency Practice or Requirement for Performing an After-Action Report	Key Contracting and Mission Assignment Findings from the 2017 Disasters	Proposed Actions
Federal Emergency Management Agency (FEMA)	Regional offices typically perform after-action reports for smaller disasters, according to FEMA officials. However, given the scope of the 2017 hurricanes, FEMA headquarters performed a combined report for Hurricanes Harvey, Irma, and Maria.	FEMA found, among other things, that the increased contracting demands from the hurricane season severely taxed FEMA's acquisitions process and contracting personnel.	None in the report. However, in September 2018, FEMA officials told us that they plan to hire 57 Stafford Act employees for 2-year appointments.[b]
Agency	Agency Practice or Requirement for Performing an After-Action Report	Key Contracting and Mission Assignment Findings from the 2017 Disasters	Proposed Actions
U.S. Army Corps of Engineers (USACE)	USACE typically performs after-action reports following a disaster, and completed several—both at the headquarters level and the regional level—following the 2017 disasters.	USACE's Northern California Wildfires Recovery Field Office found, among other things, that unpredictable staffing levels in field offices resulted in inefficient workload distribution.	USACE recommended that the agency develop guidance to transfer responsibilities from one staff office to another when work is distributed inefficiently.

Table 4. (Continued)

Agency	Agency Practice or Requirement for Performing an After-Action Report	Key Contracting and Mission Assignment Findings from the 2017 Disasters	Proposed Actions
Coast Guard	The Coast Guard is required to complete an after-action report following exercises, training events, and real-world events such as the 2017 disasters.	The Coast Guard found, among other things, that it would greatly benefit from a dedicated and highly trained team of experts to coordinate mission assignments—the process through which FEMA assigns disaster response work to other agencies.	Coast Guard officials we spoke with have since decided that more coordination could happen across locations, without the need of a dedicated support team.
Defense Logistics Agency (DLA)	The agency is required to complete an after-action report following events, such as exercises and real-world events.	DLA found, among other things, that there were several requirements during the 2017 disasters that were not included in the work outlined in the mission assignments that were prepared prior to a disaster occurring, such as the supply of bottled water.	DLA recommended that it review existing mission assignments to determine if they should be modified or if new assignments are needed to pre-plan for future disasters.

Source: GAO analysis of agency policy and after-action reports and interviews with agency officials. | GAO-19-281.

[a]To select these key findings, we reviewed agencies' after-action reports and identified findings that directly related to an agency's use of post-disaster contracts, such as workforce issues, or the mission assignment process. Two of the agencies in our scope—USACE and the Coast Guard—used post-disaster contracts to fulfill their assigned missions.

[b]FEMA's after-action report recommended that the agency accelerate ongoing efforts to increase contract support capacities but did not identify how to do so. What FEMA refers to as Stafford Act employees include a Cadre of On-Call Response/Recovery Employees who are temporary employees with 2- to 4-year appointments and can be deployed to fulfill any role specifically related to the incident for which they are hired and qualified.

Table 4 lays out each agency's practice or requirement for identifying lessons learned and key findings— those related to contracting and mission assignments during the 2017 disasters.

FEMA Established Interagency Lessons Learned Group but the Coast Guard and USACE Could Enhance Information Sharing

FEMA has also taken steps to identify interagency lessons learned by leading the Emergency Support Function Leadership Group and developing a mechanism to regularly report to the Secretary of Homeland Security.[38] This group consists of the national emergency support function coordinators from each of the functions (such as transportation and firefighting), along with FEMA headquarters and regional officials. This body of senior officials is tasked with coordinating responsibilities and resolving operational and preparedness issues relating to interagency response activities in support of the National Response Framework.[39] According to its charter, the group is required to carry out post-incident and after-exercise critiques, and perform substantive reviews of after-action reports, with recommendations for federal interagency partners to address shortfalls. Following the 2017 disasters, in May 2018, the Emergency Support Function Leadership Group identified 19 corrective actions, including improvements to mission assignment submission documents.

Federal internal control standards state that communicating internally is key to an entity achieving its objectives. Further, as part of this communication, management should receive quality information about the entity's operational processes that flows up the reporting lines from personnel to help management achieve the entity's objectives.[40] FEMA officials stated that there are processes, such as data calls, in place to solicit

[38] GAO, *Emergency Preparedness: Opportunities Exist to Strengthen Interagency Assessments and Accountability for Closing Capability Gaps*, GAO-15-20 (Washington, D.C.: Dec. 9, 2015). In December 2018, GAO closed as implemented its recommendation that FEMA, in coordination with other federal departments and agencies, collect information on federal interagency implementation of corrective actions identified through national-level exercises and following real-world incidents and regularly report the status to the Secretary of Homeland Security.

[39] At this time, the Emergency Support Function Leadership Group does include a representative from emergency support function 14, although this function is under revision. However, FEMA officials stated that emergency support function 14 is covered through other representatives.

[40] GAO-14-704G.

input from agencies. However, we noted, and FEMA officials agreed, that there is no formal reporting mechanism to the leadership group, and that it is up to the representatives from these agencies to raise issues for the group's consideration.

However, this is not consistently happening within the Coast Guard because it does not have a formal reporting process for soliciting input from officials directly involved in responding to these disasters to share with the Emergency Support Function Leadership Group. Coast Guard officials stated that they actively collect input during and immediately after an event or incident response, and that Coast Guard responders are able to provide input and issues through their chain of command at any time, but there is no formal process for reporting to the interagency group. During the course of our review, USACE officials did not provide information that indicated they had a formal reporting process for soliciting input from officials directly involved in responding to these disasters to share with the Emergency Support Function Leadership Group. Some senior level USACE officials responsible for the agency's public works and engineering mission stated that they were unsure of the process for raising concerns to the Emergency Support Function Leadership Group and that officials were sometimes hesitant to raise issues to the group. However, in response to our draft report, USACE stated it has a formal process called the USACE Remedial Action Program for soliciting input from officials directly involved in the agency's response and recovery following a disaster. As discussed later, we will follow up with USACE as part of our recommendation follow-up process. While Emergency Support Function Leadership Group member agencies may raise issues to the group, additional opportunities exist within these agencies to enhance the lines of communication from responders to the senior officials that comprise this leadership group. For example, some of the interagency challenges we identified in our review were not identified by this group, such as challenges in managing state and local expectations of federal response, which is discussed in more detail below. Also, USACE officials told us that some of the interagency challenges they cited following the 2017 disasters related to the mission assignment process were still present during the response to Hurricane Florence, which struck the Carolina coast in 2018. Formal processes for Emergency Support Function agencies—such as the Coast Guard and USACE—to solicit and share input from officials directly involved in the response and recovery efforts would help ensure the Emergency Support Function Leadership Group does not miss additional opportunities to improve disaster response.

Interagency Contracting Coordination Challenges within the Mission Assignment Process Remain

As the federal disaster coordinator, FEMA obtains requirements from states and localities and tasks the appropriate federal agencies, based on their emergency support function, through the mission assignment process. The agency assigned to a specific mission is then responsible for fulfilling those requirements, and may use contracts to do so. For example, the Coast Guard fulfills its pollution mitigation mission by executing contracts, and utilizes its own workforce to execute its search and rescue mission.

USACE officials we spoke with raised concerns about the mission assignment process for the debris removal and power restoration missions related to the 2017 disasters. Specifically, USACE officials noted concerns about coordination between state, local, and federal partners for the contracts we reviewed.

- USACE debris removal mission: In December 2018, we found that USACE and California state officials reported different expectations related to USACE's debris removal contracts following the wildfires, such as what structures would be removed from private properties and what levels of soil contamination would be acceptable.[41] USACE removed more than 2.2 million tons of debris from more than 4,500 properties following the northern California wildfires. Due to the size and scope of this

[41] GAO-19-93.

mission, USACE used both its advance contracts and additional post-disaster contracts for debris removal. According to USACE officials, they relied on FEMA, the lead for coordinating federal disaster response, to manage communication with states and localities and to identify and manage expectations about the scope of work to be performed using their debris removal contracts.

- USACE officials cited challenges with communicating to state and local officials what the agency was permitted to do under its mission assignment. For example, USACE officials told us that local officials believed that USACE would replace soil removed as part of its debris removal efforts; however, this was not part of the mission assignment from FEMA. Further, officials added that different environmental standards created confusion regarding what types of soil should be removed. For example, Napa County officials said that USACE's mission required them to ensure that no contaminated soil remained on the properties, without regard for the naturally occurring levels of arsenic and asbestos in Napa area soil. As a result, Napa County officials said that USACE removed more soil than was necessary. However, following discussions with Napa County officials, USACE obtained site-specific samples from some properties to understand pre-existing contamination levels prior to further debris removal.
- USACE power restoration mission: Hurricane Maria destroyed much of the electricity grid in Puerto Rico, leaving millions without power and resulting in the longest blackout in U.S. history. To restore power to its 3.3 million people, Puerto Rico requested federal assistance with its power grid. To coordinate this effort across all stakeholders, FEMA established a unified command structure—which included the federal agencies, the Puerto Rican government and its contractors, and utility companies providing mutual assistance. According to FEMA officials, this structure allowed stakeholders to target priority work, ensure crews could access the work areas, and identify the needed materials. USACE officials stated that they received direction from FEMA and had

limited direct interaction with Puerto Rican officials. However, despite this structure, USACE officials noted that changing direction from FEMA contributed to inefficiencies in contract management. For example, the scope of power restoration work Puerto Rico was requesting changed several times—such as from transmission work to distribution. These changes necessitated adjustments in contractor workforce configurations and contributed to idle time and equipment, according to officials.

FEMA's mission assignment policy designates a Federal Disaster Recovery Coordinator as the person responsible for facilitating disaster recovery coordination and collaboration among federal, state, local, tribal, and territorial governments; the private sector; and voluntary, faith-based, and community organizations. However, neither FEMA's mission assignment policy nor its guide—which provides guidance on how to open and close mission assignments—provide additional details on how that coordination is to take place. Further, FEMA's Response Directorate—the office that oversees the mission assignment process— was unable to identify at what level this coordination should occur.

USACE and Coast Guard officials also noted that the mission assignment process does not account for other contracting considerations, such as demobilization, which occurs when contractor personnel leave the work site and return to their headquarters. According to USACE and Coast Guard officials, demobilization is required to be completed by the end of the contract's period of performance; therefore, contracting officers need to know when the mission will end so that they can build adequate time for demobilization into the contract.

- Coast Guard pollution mitigation mission: Under this mission, the Coast Guard is responsible for responding to threats to public health, welfare, or the environment caused by actual or potential oil and hazardous materials incidents. Coast Guard officials told us that mission timing and the length of requirements were not communicated by FEMA in a timely manner. They told us that they contacted FEMA multiple times to determine if its mission assignment would be continued, but they did not receive an answer until shortly before the end of a contract's period of performance. As a result, officials told us they were unsure whether they would need to demobilize contractors before completing the work, which created uncertainty about the availability of subcontractors. A FEMA Response Directorate official stated that these issues are coordination and planning concerns that should be worked out in advance between FEMA and the mission assigned agency. Ultimately, FEMA extended the Coast Guard's mission assignment for pollution mitigation following Hurricane Maria four times. Figure 13 depicts the number of times Coast Guard's mission was extended by FEMA.

- USACE power restoration mission: USACE officials cited similar challenges during the power restoration mission in Puerto Rico following Hurricane Maria. For example, USACE officials stated they typically begin planning for demobilization as soon as a mission begins. However, in this instance, officials did not know the eventual end date in order to plan for demobilization activities. Officials added that demobilization may take about 30 days, but USACE cannot extend contracts or obligate funds without a FEMA mission assignment extension. For example, if the mission assignment is scheduled to end on June 30, contracting officials would need to direct the contractor to begin demobilization as early as May 31. Officials stated that a mission assignment extension or option period of 30 days beyond the anticipated mission end date would facilitate demobilization and reduce any undue burden or concern around demobilization efforts.

FEMA's mission assignment guide does not provide a process or mechanism to follow up on the status of a mission once it is assigned. A FEMA official stated that the Response Directorate is responsible for informing their leadership of expiring mission assignments and contacting the mission-assigned agency to make them aware of the impending expiration, but that there is no standard time frame for doing so.

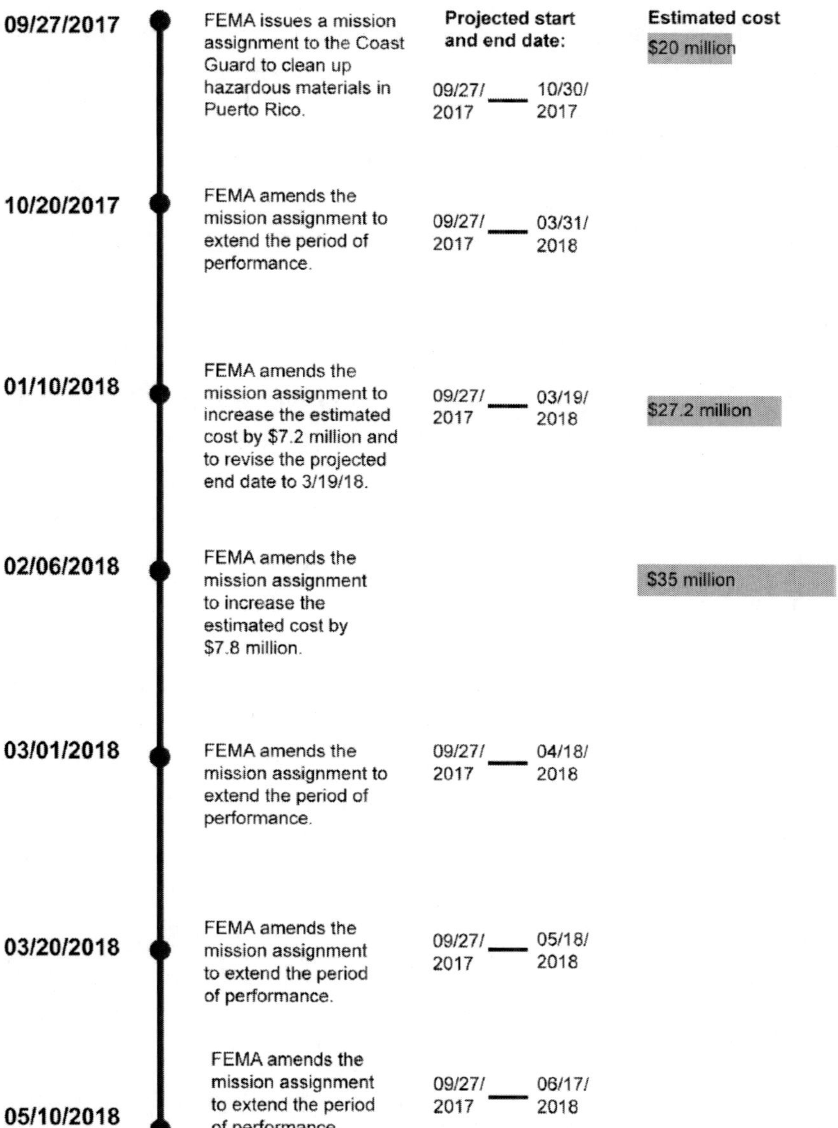

Source: GAO analysis of U.S. Coast Guard contract documentation. I GAO-19-281.

Figure 13. Timeline of Coast Guard Mission Assignment for Pollution Mitigation for Hurricane Maria.

Further, the official stated that, in some cases, FEMA may be performing this work a few days before a mission is set to expire. However, officials at USACE and Coast Guard told us they are dependent upon FEMA to reissue, clarify, or extend mission assignments. Further, the FEMA official told us that contracting considerations—such as the time needed for a contractor to mobilize and demobilize—are not necessarily built into the period of performance of a mission assignment.

FEMA identified issues related to the mission assignment process, both during the 2017 disasters and following Hurricane Sandy in 2012. For example, in its 2013 Hurricane Sandy After-Action Report, FEMA found that the mission assignment process was not optimally set up to quickly surge resources to the field in a large-scale incident. To address these challenges, FEMA convened an Executive Steering Committee to update the mission assignment process, among other actions, and subsequently updated its mission assignment policy in 2015. Following the 2017 disasters, the Emergency Support Function Leadership Group identified challenges related to the mission assignment process and made recommendations to:

1) ensure response officials are properly trained on their department or agency's statutory authorities and FEMA's mission assignment process, and
2) develop specific recommendations to the FEMA Response Directorate on ways to reform mission assignment submission documents.

These recommendations have been assigned to working groups within the Emergency Support Function Leadership Group, which plans to track the status until they are implemented. While these actions may improve the mission assignment process, they do not specifically address the issues we identified related to coordination and contracting.

While the emergency support functions lay out agencies' general responsibilities, agencies are dependent upon FEMA's mission assignment process to further define how to perform their roles. Federal internal control standards state that management should implement control activities through its policies. These control activities include periodically reviewing policies, procedures, and related control activities for continued relevance and effectiveness in achieving the entity's objectives or addressing related risks. Further, these standards also state that communicating internally and externally are key to achieving an entity's objectives. As part of its internal controls, entities should evaluate the methods to communicate quality information throughout and outside of the entity on a timely basis.[42] While FEMA revised its mission assignment guide in 2017, it still does not require FEMA to lay out coordination responsibilities in detail when assigning a mission. Without a mission assignment policy and related guidance that better incorporates contracting considerations, such as demobilization, and requires FEMA to clearly define coordination responsibilities with federal, state, and local stakeholders during the mission assignment process, federal agencies may encounter challenges fulfilling their assigned missions and may not fulfill their disaster response and recovery missions efficiently.

FEMA Identified Contracting Workforce Shortages, but Has Not Fully Assessed Its Needs

During the 2017 disasters, FEMA leveraged contracting staff from its regions, headquarters, and the DART teams—FEMA's deployable contracting workforce. However, FEMA's after-action report and officials we spoke with cited workforce shortages as a continuing challenge for disaster response and recovery. For example, officials we spoke with in several regional offices stated that there are only one to three contracting officers per region. Further, information provided by FEMA OCPO shows that eight of FEMA's 10 regional offices have only one permanent full-time contracting official. Some of FEMA's regional offices have additional contracting staff through FEMA's Cadre of On-Call Response/Recovery

---

[42] GAO-14-704G.

Employees, but this varies from region to region.[43] Regional offices are responsible for managing post-disaster contracts, even if regional procurement staff were not involved in the initial award of those contracts, according to FEMA officials.

As noted in Table 4 above, FEMA's after-action report recommended increasing contract support capacities; however, it did not provide a specific plan to do so. According to FEMA officials, the agency's workforce needs have not been assessed since a FEMA workforce analysis pilot conducted in 2014. We have identified several key principles that strategic workforce planning should address, including:

- determining the critical skills and competencies that will be needed to achieve current and future programmatic results, and
- developing strategies that are tailored to address gaps in the number, deployment, and alignment of human capital approaches for enabling and sustaining the contributions of all critical skills and competencies.

Further, in our review of FEMA's 2014 analysis, we found that FEMA evaluated contracting workforce needs, but did not specifically consider contracting workforce needs in the regional offices or address DART employees. The analysis was based on 5 years of workload data and conducted at the task or activity level, such as performing market research prior to making a contract award. However, the analysis did not prioritize skills or mission needs, nor did it identify critical competencies. In September 2018, FEMA procurement officials told us that, based on the 2014 analysis, they planned to hire 57 additional contracting staff. Officials noted that FEMA's general operation funding does not support these additional hires, thus the agency plans to hire these staff as Stafford Act employees for 2-year appointments using disaster funding.[44] While this is

[43] FEMA's Cadre of On-Call Response/Recovery Employees are temporary employees with 2- to 4-year appointments who can be deployed to fulfill any role specifically related to the incident for which they are hired and qualified.

[44] What FEMA refers to as Stafford Act employees includes its Cadre of On-Call Response/Recovery Employees.

an important step, it is unclear when these staff will be hired or how they will be allocated across FEMA OCPO. For example, as of July 2018, FEMA OCPO had 72 vacant positions, including key leadership positions and contracting specialists. Without assessing its current contracting workforce needs—including staffing levels, mission needs, and skill gaps—and developing a plan to address these gaps that includes time frames, FEMA will not know whether it has the appropriate number of contracting officials with the key skills needed to meet its mission and is not likely to be well-positioned to respond to future disasters.

CONCLUSION

Contracting during a disaster can pose a unique set of challenges as officials face a significant amount of pressure to provide life-sustaining goods and services to survivors as quickly as possible. Given the scale and consecutive nature of the 2017 disasters, disaster contracts— particularly post-disaster contracts—played a key role in the response and recovery efforts. In these situations, it is important that the federal government be accountable for the contracting decisions it makes and the money it obligates, support the local economy and survivors as effectively as possible, and implement lessons learned before the next disaster strikes.

Regarding accountability for the contracting decisions it makes and dollars obligated following disasters, without the ability to track disaster contracts using a NIA code in FPDS-NG, agencies, Congress, and the public lack full insight into post-disaster contracts. Providing clear criteria for establishing and closing the NIA code that accounts for the needs of users and consistently implementing these criteria will help ensure insight into high-visibility disaster events. Further, the ability to identify and track contracting dollars for disasters through a publicly available database, such as FPDS-NG, can reduce the burden on agencies to provide these data for interested parties, including Congress and other users, and offer a resource for historical data across major disasters.

To help meet the needs of the local economy as effectively as possible, using a contracting preference for vendors in a disaster-affected area is an important component to early recovery efforts. Without guidance or training to ensure contracting officers are aware of the regulatory definition of the local area, agencies may miss opportunities to provide financial support to local vendors. Additionally, without clarifying how contracting officers determine whether offerors reside or primarily do business in a disaster area for the purposes of a local area set-aside, contract officials will remain uncertain on how to implement related FAR criteria. Similarly, guidance and tools to help ensure contracting officials are aware of the requirement to provide preference to the extent feasible and practicable to local vendors, including the need for written documentation on the use of non-local vendors for post-disaster contracts, will help ensure agencies comply with the requirement to do so. Taken together, these actions could enhance compliance with the Stafford Act provisions related to the award of contracts to local businesses in the disaster area, which could help jump-start the local economy.

With regards to implementing lessons learned before the next disaster strikes, large scale disasters, like those that occurred in 2017, require effective coordination across emergency support function agencies. Given the Emergency Support Function Leadership Group's responsibility to identify gaps or seams in the federal government's efforts to respond to disasters, it is essential that the group have accurate and up-to-date information. Formal processes for soliciting and sharing information to communicate lessons learned to this group would help enhance agencies' abilities to identify and address weaknesses in disaster response. Further, incorporating contracting considerations, such as demobilization, into the mission assignment policy, could enhance federal agencies' ability to fulfill their disaster response and recovery missions efficiently. Lastly, without an assessment of FEMA's contracting workforce needs, FEMA is at risk of not having a sufficient contracting workforce during a disaster.

RECOMMENDATIONS FOR EXECUTIVE ACTION

We are making a total of 10 recommendations, including one to DHS, one to the Office of Federal Procurement Policy, two to FEMA, three to the Army, two to the Coast Guard, and one to GSA (in coordination with DOD and DHS).

The Administrator of the General Services Administration, in coordination with the Secretaries of Defense and Homeland Security, should jointly revisit and assess the extent to which the criteria in the 2018 NIA code Memorandum of Agreement, including criteria for closing NIA codes, meet long-term visibility needs for high visibility events and account for the needs of users, such as FEMA, other agencies, and the Congress. At a minimum, the agreement should include criteria that take into account the roles of the federal agencies involved in response and recovery and provide a process that ensures consistent consideration and implementation of the criteria. (Recommendation 1)

Until the NIA code Memorandum of Agreement between the General Services Administration and the Departments of Defense and Homeland Security is revised, the Secretary of Homeland Security should, in coordination with the Department of Defense and the General Services Administration, keep the existing NIA code for Hurricane Maria open, reopen the other NIA codes established for 2017 and 2018 hurricanes (Hurricanes Harvey, Irma, Florence, and Michael), and request that agencies retroactively enter NIA codes for contract actions for Hurricanes Harvey and Irma made after June 30, 2018, for Hurricane Florence made after March 15, 2019, and for Hurricane Michael made after April 12, 2019 into FPDS-NG to adequately capture contract obligations, to the extent practicable. (Recommendation 2)

The Secretary of the Army should direct the Commanding General of the U.S. Army Corps of Engineers to provide guidance or related training to ensure contracting officers are aware of the regulatory definition of "local area." (Recommendation 3)

The Administrator of the Office of Federal Procurement Policy should provide additional clarification on how contracting officers should determine whether offerors reside or primarily do business in a disaster area for the purposes of a local area set-aside contract. (Recommendation 4)

The Commandant of the Coast Guard should provide guidance and tools for contracting officials to use to ensure requirements concerning contracting with local vendors, including justification requirements for the use of non-local vendors, are consistently met. (Recommendation 5)

The Secretary of the Army should direct the Commanding General of the U.S. Army Corps of Engineers to provide guidance and tools for contracting officials to use to ensure requirements concerning contracting with local vendors, including justification requirements for the use of non-local vendors, are consistently met. (Recommendation 6)

The Secretary of the Army should direct the Commanding General of the U.S. Army Corps of Engineers to establish a formal process to solicit input from officials directly involved in the agency's response and recovery following a disaster and to share that input with the Emergency Support Function Leadership Group. (Recommendation 7)

The Commandant of the Coast Guard should establish a formal process to solicit input from officials directly involved in the agency's response and recovery following a disaster and to share that input with the Emergency Support Function Leadership Group. (Recommendation 8)

The FEMA Administrator should take the lead to work together with the Coast Guard and the U.S. Army Corps of Engineers to revise the mission assignment policy and related guidance to better incorporate consideration of contracting needs, such as demobilization, and to ensure clear communication of coordination responsibilities related to contracting. (Recommendation 9)

The FEMA Administrator should assess its workforce needs—including staffing levels, mission needs, and skill gaps—for contracting staff, to include regional offices and DART; and develop a plan, including timelines, to address any gaps. (Recommendation 10)

AGENCY COMMENTS AND OUR EVALUATION

We provided a draft of this chapter to DOD, DHS, GSA, and OMB for review and comment. In written comments provided by DOD, DHS, and GSA (reproduced in appendixes III, IV, and V), as well as an email response from OMB, the agencies concurred with nine of the 10 recommendations. They generally provided steps they plan to take to address these recommendations. As discussed further below, USACE described actions it stated were sufficient to fully address the seventh recommendation, the steps described by FEMA would not fully meet the intent of the tenth recommendation, and DHS did not concur with our second recommendation.

In response to the seventh recommendation as written in our draft report—to establish a formal process to solicit input from officials directly involved in the agency's response and recovery following a disaster and to share that input with the Emergency Support Function Leadership Group—in its comments, USACE concurred and stated it has a formal process and it considered the recommendation completed. USACE noted that its Remedial Action Program solicits input from officials involved in response and recovery efforts and added that USACE shares findings from this program with the Emergency Support Function Leadership Group throughout the year and annually during the senior leaders seminar. During the course of our review, USACE did not provide information that indicated that they had such a formal process. As part of our recommendation follow-up process, we will request documentation regarding the process and how it solicits and shares information to the Emergency Support Function Leadership Group.

In response to the tenth recommendation that FEMA assess its workforce needs—including staffing levels, mission needs, and skill gaps—for contracting staff, to include regional offices and DART; and develop a plan, including timelines, to address any gaps, FEMA stated that its Office of the Chief Component Procurement Officer assesses its workforce on an annual basis, with the last assessment conducted in January 2019. FEMA also noted that it entered into a contract for acquisition support services and plans to hire Cadre of On-Call Response and Recovery employees to provide dedicated support during disasters. Following FEMA's response, we requested and received the FEMA Office of the Chief Component Procurement Officer's 2019 workforce assessment. As with FEMA's 2014 workforce analysis, the 2019 assessment calculated the number of employees needed based on the estimated time to complete a task. However, the assessment did not include an analysis of mission needs or skill gaps, and the assessment provided does not specify whether it includes the needs of regional offices and DART. FEMA estimates that it will implement this recommendation in September 2019, and we will continue to monitor FEMA's planned efforts through our recommendation follow-up process.

DHS did not concur with the draft report's second recommendation regarding NIA codes. In its response, with regards to extending existing NIA codes and reinstating expired NIA codes, DHS stated that it is bound by the memorandum of agreement with GSA and DOD, unless or until all three signatory agencies agree to revise or suspend the agreement. We recognize that all three agencies are bound by the agreement, and also recommended in the first recommendation that GSA, DOD, and DHS jointly revisit the agreement. GSA concurred with this recommendation in its written comments reproduced in Appendix V. In an email sent from an official within DOD's Defense Pricing and Contracting Office, DOD concurred. DHS did not respond to our first recommendation. As such, we have revised the second recommendation to state that DHS take action in coordination with DOD and GSA. We also note that the memorandum of agreement states that extending expiring or already expired NIA code end date is appropriate, in part, when two or more agencies do not have a reasonable alternative method of identifying and internally tracking those emergency acquisitions. We discuss in our report how once the NIA code is closed, there is no publicly available, government-wide system to track contract obligations for specific events. We also discuss how, using the description field (which does not provide a full picture) in FPDS-NG, agencies obligated more than $250 million on contracts for Hurricanes Harvey and Irma during the three months after the NIA codes for these two hurricanes were closed. Given this, we continue to believe DHS should consider reopening the codes for Hurricanes Harvey and Irma, in coordination with DOD and GSA.

Moreover, in its response to the second recommendation DHS further stated that FEMA's Office of the Chief Component Procurement Officer (who is not currently a party to the memorandum of agreement), believes the recommendation to extend the NIA codes for 2018 Hurricanes Michael and Florence goes beyond the scope of this audit. While the main focus of this chapter is the 2017 hurricanes and California wildfires, we discuss Hurricanes Florence and Michael in this draft with respect to the NIA codes, as the same issues and concerns we raised apply regardless of the year of the hurricane. However, after we sent the draft to the agencies for comment, the agencies let the codes for Hurricanes Florence and Michael expire on March 15, 2019 and April 12, 2019, respectively. We therefore revised the second recommendation to recommend that the codes for Hurricanes Florence and Michael should be reopened (rather than kept open).

In its written comments, DHS also stated that neither DHS nor FEMA can unilaterally direct other agencies to retroactively enter FPDS-NG data for Hurricanes Harvey and Irma. We acknowledge this and have revised the recommendation to recommend that DHS request, rather than direct, other agencies to retroactively enter the information, to the extent practicable. As we state in the report, the NIA codes for the 2005 hurricanes were established in October 2005, and contracting officers retroactively entered data for contracts related to these events to enable full insight into contracting for these disasters. DHS further stated that retroactively entering data into FPDS-NG is not practical and places an unreasonable burden on contracting staff, and that the draft did not support the case that there were any benefits to be gained. We recognize that there is some burden associated with the recommendation, thus we recommended that DHS request agencies take action to the extent practicable. In terms of benefits, the report identifies benefits in terms of providing decision makers with important information to understand the procurement impact of such disasters.

DOD and DHS also provided technical comments, which we incorporated as appropriate.

We are sending copies of this chapter to the appropriate congressional committees, the Secretary of Defense, the U.S. Army Corps of Engineers Director of Contracting, the Director of the Defense Logistics Agency, the Secretary of Homeland Security, the Administrator of the Federal Emergency Management Agency, the Federal Emergency Management Agency's Chief Procurement Officer, the Commandant of the Coast Guard, the Administrator of the General Services Administration, the Director of the Office of Management and Budget, and the Administrator of the Office of Federal Procurement Policy.

Marie A. Mak
Director, Contracting and National Security Acquisitions

List of Requesters

The Honorable Michael B. Enzi
Chairman
Committee on the Budget
United States Senate

The Honorable Ron Johnson
Chairman

The Honorable Gary C. Peters
Ranking Member
Committee on Homeland Security and Governmental Affairs
United States Senate

The Honorable Rand Paul, MD
Chairman
Subcommittee on Federal Spending Oversight and Emergency Management
Committee on Homeland Security and Governmental Affairs
United States Senate

The Honorable Marco Rubio
Chairman
Committee on Small Business and Entrepreneurship
United States Senate

The Honorable Bennie G. Thompson
Chairman
Committee on Homeland Security
House of Representatives

The Honorable Elijah E. Cummings
Chairman

The Honorable Jim Jordan
Ranking Member
Committee on Oversight and Reform
House of Representatives

The Honorable Peter DeFazio
Chairman

The Honorable Sam Graves
Ranking Member
Committee on Transportation and Infrastructure
House of Representatives

The Honorable Maxine Waters
Chairwoman
Committee on Financial Services
House of Representatives

The Honorable Nydia M. Velazquez
Chairwoman
Committee on Small Business
House of Representatives

The Honorable Al Green
Chairman
Subcommittee on Oversight and Investigations
Committee on Financial Services
House of Representatives

The Honorable Sean Duffy
Ranking Member
Subcommittee on Housing, Community Development and Insurance
Committee on Financial Services
House of Representatives

The Honorable Emanuel Cleaver, II
House of Representatives

The Honorable Michael McCaul
House of Representatives

The Honorable Gary J. Palmer
House of Representatives

The Honorable Ann Wagner
House of Representatives

APPENDIX I: OBJECTIVES, SCOPE, AND METHODOLOGY

This chapter specifically addresses the use of post disaster contracts and: (1) assesses the extent to which federal agencies obligated funds on post-disaster contracts in response to the 2017 major disasters; (2) assesses the extent to which selected agencies experienced challenges in the planning process for selected post-disaster contracts; and (3) describes selected agencies' lessons learned as a result of the 2017 major disasters and assesses the extent to which they have taken action to address them.

To identify the extent to which federal agencies obligated funds on post-disaster contracts in response to the 2017 disasters, we reviewed Federal Procurement Data System-Next Generation (FPDS-NG) data through June 30, 2018, the most recent and complete data at the time of our review.[45] We adjusted the obligation data to constant fiscal year 2018 dollars using the Fiscal Year Gross Domestic Product price index. We identified hurricane obligations using the national interest action (NIA) code, as well as the contract description.

Data on obligations for the California wildfires is limited to those contracts, if any, identified by the agencies with the highest obligations on post-disaster contracts for the hurricanes—the Federal Emergency Management Agency (FEMA), U.S. Army Corps of Engineers (USACE), Defense Logistics Agency (DLA), and the U.S. Coast Guard (Coast Guard)—because no NIA code was established in FPDS-NG.[46] Coast Guard officials stated that they did not execute any contracts in response to

---

[45] For the purposes of this chapter, contract obligations include obligations against what the General Services Administration's FPDS-NG categorizes as definitive vehicles (definitive contracts and purchase orders that have a defined scope of work that do not allow for individual orders under them), and against what FPDS-NG categorizes as indefinite delivery vehicles (orders under the Federal Supply Schedule, orders/calls under blanket purchase agreements, orders under basic ordering agreements, orders under government-wide acquisition contracts, and orders under other indefinite delivery vehicles, such as indefinite delivery, indefinite quantity contracts).

[46] NIA codes are used in FPDS-NG to track all awards, indefinite delivery vehicles and modifications for both civilian agencies and DOD used for disaster response and recovery. No NIA code was requested for the 2017 California wildfires. Post-disaster contract obligations for the California wildfires reflect agency identified obligations for Disaster Response 4344 and Disaster Response 4353.

the 2017 California wildfires. DLA officials stated that they maintain contracts, which for the most part provide inventory replenishment for DLA and the U.S. Forest Service within the U.S. Department of Agriculture, but they were unable to provide data on contracts awarded or executed specifically for the two wildfire disasters in the scope of our review. Therefore, our analysis only captures obligations for FEMA and USACE reported contracts related to the 2017 California wildfires.

To determine which obligations were made through the use of post-disaster contracts versus advance contracts, we reviewed documentation provided by FEMA and USACE identifying the advance contracts they have in place and that were used in support of the 2017 disasters. We analyzed the FPDS-NG data against these contracts to identify obligations on post-disaster contracts and compared these to obligations on advance contracts by disaster. We analyzed competition procedures used and the types of goods and services procured for post-disaster contracts. In addition to advance contracts for disaster response, agencies can leverage other existing contract vehicles. For example, to respond to its pollution mitigation functions under emergency support function 10, the Coast Guard awards task orders off of its portfolio of basic ordering agreements. For the purposes of this chapter, post-disaster contracts include all contract awards and orders that were not identified by FEMA or USACE as advance contracts.

To assess the extent to which disaster contract obligations can be tracked through FPDS-NG using the NIA code, we identified prior hurricane events with the highest contract obligations from 2005 through September 2018. We analyzed the data to determine when the highest level of federal contract obligations occurs following a hurricane. We also assessed the process for establishing and closing a NIA code. Specifically, we reviewed the criteria in the 2012 and 2018 memorandums of agreement between DHS, DOD, and the General Services Administration, and interviewed officials involved in the process.

We assessed the reliability of FPDS-NG data by reviewing existing information about the FPDS-NG system and the data it collects—specifically, the data dictionary and data validation rules—and performing electronic testing. We also compared FPDS-NG data to the contract files in our review. Specifically, to review our selected post-disaster contracts for data reliability, we compared items such as, the extent competed, the use of a local area set-aside, NIA code, and termination status, based on the contract information and the information in FPDS-NG. Based on the steps we took, we determined the FPDS-NG data were sufficiently reliable for the purposes of describing agencies' post- disaster contract obligations.

To assess the extent to which agencies experienced challenges in the planning of selected post-disaster contracts, we reviewed relevant laws and regulations, including the Post-Katrina Emergency Management Reform Act (PKEMRA), the Federal Acquisition Regulation (FAR), the Robert T. Stafford Disaster Relief and Emergency Assistance Act (Stafford Act), as well as agency policy and guidance. We identified a non-generalizable sample of 23 post-disaster contracts from the four agencies with the highest post-disaster obligations based on FPDS-NG data as of March 31, 2018—DHS's FEMA, DOD's USACE, DOD's DLA, and DHS's Coast Guard. We selected contracts across the four major 2017 disasters included in our scope (Hurricanes Harvey, Irma, and Maria, as well as the California wildfires) based on four selection criteria—(1) contracts using the urgency exception to full and open competition; (2) contracts using a local area set-aside; (3) contracts awarded to small businesses; and (4) contracts terminated for cause or convenience.[47] Our goal in this selection was to ensure we selected a range of contracts within each of these four criteria so as to assess the extent to which these contracts implemented certain laws and regulations. Specifically, we selected contracts based on the use of urgency and local area set-asides in order to assess agencies' implementation of relevant PKEMRA, Stafford Act and FAR criteria for post-disaster contracts. Because the obligations for local area set-aside contracts was low across all federal agencies, about 5 percent of total post-

[47] Termination for Convenience is the government's unilateral contractual right to partially or completely terminate a contract. For example, a Contracting Officer may terminate for convenience when the requirement is no longer needed. When terminating for convenience the government is required to come to a settlement that compensates the contractor for the work done and preparations made for the terminated portion of the contract. The government may terminate all or a portion of a commercial item contract for cause if the contractor fails to provide items or services that conform to the contract requirements.

disaster obligations, we selected contracts that were awarded to small business vendors as a proxy to identify other awards to local vendors. Finally, we selected terminated contracts to assess additional challenges related to post-disaster contracts, such as the availability of contracted services and supplies and the requirement setting process. Based on these criteria, we selected 12 FEMA, 7 USACE, 2 DLA, and 2 Coast Guard contracts. Findings based on information collected from the 23 contracts cannot be generalized to all post-disaster contracts. Additional details on our selected contracts can be found in Table 5.

To assess how agencies used the urgency exception to full and open competition, we reviewed selected contracts for the inclusion of a justification and approval for other than full and open competition including sole source justifications and exclusion of sources justifications.

To assess the extent to which agencies provided preference to local vendors for post-disaster contracts, we reviewed selected contract files for the use of a set-aside or an evaluation preference listed in the contract solicitation, and the inclusion of justifications for contracts not awarded to local vendors. Additionally, we reviewed applicable agency guidance and interviewed contracting and senior procurement officials across all four agencies regarding their use of local area set-asides, including the means by which they define the geographic set-aside area and determine that an offeror primarily resides or does business in the set-aside area. We also met with officials from the Office of Management and Budget's Office of Federal Procurement Policy to discuss relevant FAR criteria.

To assess how FEMA program offices develop and deliver requirements packages for use by contracting officers and the extent to which those packages are sufficiently specific to allow contracting officers to issue a contract solicitation, we interviewed contracting, program, and senior procurement officials responsible for the contracts in our selection sample. We discussed the specificity of initial versus final requirements, the nature of requirements changes, the process of requirements development, and training provided to program officials regarding the requirements development process.

Table 5. Selected Post-Disaster Contracts

Description of Action	Selection Criteria Met	Event Supported		Obligations as of June 30, 2018 (in millions)
Federal Emergency Management Agency	Tank and pump systems	Urgency Exception, Small Business Vendor	Hurricane Harvey	$22.9
	Distribution management service, including warehousing, outdoor yard, transportation services, refrigerated containers.	Urgency Exception	Hurricane Maria	$21.9
	Food to supply food bank	Local Area Set-Aside, Small Business Vendor	Hurricane Harvey	$23.0
	Meals	Local Area Set-Aside	Hurricane Maria	$15.1
	Leasing or rental of equipment for Disaster Recovery Centers	Local Area Set-Aside, Small Business Vendor	Hurricane Maria	$6.5
	Food to supply food bank	Local Area Set-Aside	Hurricane Harvey	$2.8
	Travel trailers	Local Area Set-Aside, Small Business Vendor	Hurricane Irma	$3.4
	Food to supply food bank	Small Business Vendor, Local Area Set-Aside	Hurricane Harvey	$37.1
	Self-help tarps	Small Business Vendor, Terminated	Hurricane Maria	$0 (Awarded for $33.9 million, terminated for convenience; de-obligated all funds in December2017)
	Meals	Small Business Vendor	Hurricane Maria	$30.3
	Meals	Terminated	Hurricane Maria	$70.0 (Awarded for $156 million, terminated for cause; de-obligated $86 million in April 2018)

Table 5. (Continued)

Description of Action	Selection Criteria Met	Event Supported		Obligations as of June 30, 2018 (in millions)
	Self-help tarps	Terminated, Small Business Vendor	Hurricane Maria	$0 (Awarded for $21.2 million, terminated for convenience; de-obligated all funds in November 2017)
U.S. Coast Guard	Vessel removal and oil pollution mitigation / other	Urgency Exception	Hurricane Irma	$6.3
	Vessel removal and oil pollution mitigation / other	Urgency Exception, Small Business Vendor	Hurricanes Maria & Irma	$22.0
U.S. Army Corps of Engineers	Repair and restore the Puerto Rican power grid	Urgency Exception	Hurricane Maria	$999.9
	Detailed survey and report to determine the destruction and plan for repair of the power grid; line and sub-transmission repair work	Urgency Exception	Hurricanes Maria and Irma	$523.0
	Provide temporary roofing	Urgency Exception, Local Area Set-Aside, Small Business Vendor	Hurricane Maria	$34.1
	Provide temporary roofing	Small Business Vendor, Local Area Set-Aside	Hurricane Maria	$28.0
	Quality assurance services	Small Business Vendor	Hurricane Maria	$8.0
	Debris removal	Urgency Exception	California wildfires	$156.6
	Debris removal	Urgency Exception	California wildfires	$18.5
Defense Logistics Agency	Galvanized steel poles	Urgency Exception	Hurricane Maria	$32.0
	Galvanized steel poles	Urgency Exception, Small Business Vendor	Hurricane Maria	$19.6

Source: GAO review of selected contracts and Federal Procurement Data System-Next Generation. | GAO-19-281.

We also reviewed new post-disaster awards at FEMA to determine time frames between resource request to award on average for post-disaster contracts. We compared these findings to relevant agency guidance on acquisition planning. To describe lessons learned selected agencies identified related to the use of post-disaster contracts and assess the extent to which agencies have taken action to address them, we reviewed available completed after-action reports from the 2017 and prior disasters, including the *Hurricane Sandy FEMA After-Action Report, the 2017 Hurricane Season FEMA After-Action Report, USACE's Temporary Emergency Power Mission After Action Review for Hurricane Matthew, USACE's Puerto Rico After Action Review*, USACE's *Northern California Wildfires Debris Removal Mission After Action Review, the Coast Guard's 2017 Hurricane Season Strategic Lessons Learned After Action Report*, and the Defense Logistics Agency's 2017 *Hurricane After Action Meeting* papers. We also reviewed findings from the Emergency Support Function Leadership Group related to interagency lessons learned. As part of our review, we identified requirements for agencies to document or practices agencies use to document lessons learned following a disaster, agency specific and interagency lessons learned specific to post-disaster contracts and mission assignments, and recommendations or actions planned by the agencies to address them. We reviewed federal internal control standards and the Emergency Support Function Leadership Group charter and the standard operating procedures for its Preparedness Evaluation/Corrective Action Working Group.[48]

To describe challenges related to coordination with state and local officials on the use of post-disaster contracts, we interviewed FEMA, USACE, DLA, and Coast Guard officials. To obtain perspectives and examples from state and local government officials involved in disaster response, we interviewed officials in California on the use of federal contracts. We also met with state and local officials in Texas, Florida, Puerto Rico, and the U.S. Virgin Islands to discuss the federal response to

---

[48] GAO-14-704G.

the 2017 hurricanes more broadly. The information gathered from these officials is not generalizable to all officials.

To describe challenges related to the mission assignment process, we interviewed FEMA, USACE, and Coast Guard officials, including officials from FEMA's Response Directorate and the contracting officials from USACE and the Coast Guard that awarded the contracts these agencies used to fulfill their missions. We also reviewed the mission assignment documents, where FEMA assigned USACE and Coast Guard missions and laid out their responsibilities.

To assess workforce challenges, we reviewed DHS's 2014 workforce assessment, which identified gaps in FEMA's contracting workforce. We also obtained information from FEMA on its current contracting workforce in headquarters, regional offices, Disaster Assistance Response Team, and joint field offices. We also interviewed FEMA contracting officials to obtain their perspectives and experiences during the 2017 disaster season.

We conducted this performance audit from March 2018 to April 2019 in accordance with generally accepted government auditing standards. Those standards require that we plan and perform the audit to obtain sufficient, appropriate evidence to provide a reasonable basis for our findings and conclusions based on our audit objectives. We believe that the evidence obtained provides a reasonable basis for our findings and conclusions based on our audit objectives.

APPENDIX II: EMERGENCY SUPPORT FUNCTIONS RESPONSIBILITIES ACROSS THE FEDERAL GOVERNMENT

The National Response Framework identifies 14 emergency support functions (ESF) and designates a federal department or agency as the coordinating agency for each function. ESFs are the federal government's primary coordinating structure for response, and under this structure, the Federal Emergency Management Agency (FEMA) acts as the federal coordinating agency.

Table 6. Emergency Support Functions (ESF) Responsibilities across the Federal Government

ESF	Scope	Coordinator/Primary Agency	Number of Support Agencies/Offices (including those in our review)
1. Transportation·	• Transportation modes management Transportation safety • and control Stabilization and reestablishment of transportation infrastructure • ·Movement restrictions • ·Damage and impact assessment	·Department of Transportation (Ca/Pb)	15 (U.S. Army Corps of Engineers (USACE), Federal Emergency Management Agency (FEMA), Coast Guard)
2. Commun-ications	• ·Coordination with telecommunications and information technology industries • ·Reestablishment and repair of telecommunications infrastructure • ·Protection, restoration, and sustainment of national cyber and information technology resources • Oversight of communications within the Federal incident management and response structures	National Protection and Programs Directorate, Office of Cybersecurity and Communications (C/P); FEMA (P)	6
3. Public Works and Engineering	• ·Infrastructure protection and emergency repair • ·Critical infrastructure reestablishment • ·Engineering services and construction management • ·	USACE (C/P)	18(Coast Guard)

Table 6. (Continued)

ESF	Scope	Coordinator/Primary Agency	Number of Support Agencies/Offices (including those in our review)
3. Public Works and Engineering	• Emergency contracting support for life-saving and life-sustaining services		
4. Firefighting	• ·Coordination of firefighting activities • ·Support to wild land, rural, and urban firefighting operations	Forest Service and U.S. Fire Administration (C/P)	8 (USACE, Coast Guard)
5. Information and Planning	• ·Incident action planning Information collection, analysis, and dissemination	FEMA (C/P)	1
6. Mass Care, Emergency Assistance, Temporary Housing, and Human Services	Mass care Emergency assistance Disaster housing Human services	FEMA (C/P)	22(USACE)
7. Logistics	Comprehensive, national incident logistics planning, management, and sustainment capability Resource support (facility space, office equipment and supplies, contracting services, etc.)	General Services Administration and FEMA (C/P)	15 (USACE)
8. Public Health and Medical Services	Public health Medical surge support Mental health services Mass fatality management	Department of Health and Human Services (C/P)	23 (USACE, FEMA)
9. Search and Rescue	Life-saving assistance Search and rescue operations	FEMA (C/P), Coast Guard, National Parks Service, Department of Defense (P)	15 (USACE)

ESF	Scope	Coordinator/Primary Agency	Number of Support Agencies/Offices (including those in our review)
10. Oil and Hazardous Materials Response	Environmental assessment of the nature and extent of oil and hazardous materials contamination Environmental decontamination and cleanup	Environmental Protection Agency (C/P), Coast Guard (P)	21 (USACE, FEMA)
11. Agriculture and National Resources	Nutrition assistance Animal and agricultural health issue response Meat, poultry, and processed egg products safety and defense Technical expertise, coordination, and support of animal and agricultural emergency management Natural and cultural resources and historic properties protection	Department of Agriculture (C/P), Department of the Interior (P)	17 (USACE, FEMA)
12. Energy	Energy infrastructure assessment, Energy industry utilities coordination repair, and restoration Energy forecast	Department of Energy (C/P)	26 (Defense Logistics Agency, USACE, FEMA, Coast Guard)
13. Public Safety and Security	Facility and resource security Security planning and technical resource assistance Public safety and security support Support to access, traffic, and crowd control	Bureau of Alcohol, Tobacco, Firearms, and Explosives(C/P)	6
14. This emergency support function is under revision, completion expected in mid-2019.			
15. External Affairs	Public, intergovernmental, and congressional affairs Private sector outreach Community relations	DHS (C), FEMA (P)	8 (Coast Guard)

Source: GAO analysis of Emergency Support Function Annexes | GAO-19-281.

[a] ESF Coordinator.

[b] Primary Agency.

Appendix III: Comments from the Department of Defense

DEPARTMENT OF THE ARMY
OFFICE OF THE ASSISTANT SECRETARY OF THE ARMY
ACQUISITION LOGISTICS AND TECHNOLOGY
103 ARMY PENTAGON
WASHINGTON DC 20310-0103

SAAL-ZP

MEMORANDUM FOR U.S. GOVERNMENT ACCOUNTABILITY OFFICE (GAO), ATTN: DIRECTOR, MS. MARIE MAK, CONTRACTING AND NATIONAL SECURITY ACQUISITIONS, 441 G STREET, N.W. WASHINGTON, D.C.

SUBJECT: GAO Draft Report: 2017 Disaster Contracting, Actions Needed to Improve the Use of Post Disaster Contracts to Support Response and Recovery (GAO-19-281)

1. On behalf of the Acting Assistant Secretary of the Army (Acquisition, Logistics and Technology), the Office of the Deputy Assistant Secretary of the Army (Procurement) reviewed the subject draft report and I am providing the official Army position on the subject report.

2. After reviewing the audit draft report, I concur with recommendations 3, 6, and 7. The United States Army Corps of Engineers is working to close the recommendations. Detailed response and draft Alert #19-001 enclosed.

HAZLETT.STUART.A.1230146817 Digitally signed by HAZLETT.STUART.A.1230146817 Date: 2019.03.15 10:59:45 -04'00'

Encl

Stuart A. Hazlett
Deputy Assistant Secretary
of the Army (Procurement)

GAO DRAFT REPORT DATED FEBRUARY 15, 2019
GAO-19-281 (GAO CODE 102659)

"2017 DISASTER CONTRACTING: ACTIONS NEEDED TO IMPROVE THE USE OF POST-DISASTER CONTRACTS TO SUPPORT RESPONSE AND RECOVERY"

United States Army Corps of Engineers (USACE) Response
TO THE GAO RECOMMENDATIONS

RECOMMENDATION 3: The Secretary of the Army should direct the Chief of Engineers under the Assistant Secretary of the Army for Civil Works and the Commanding General of the U.S. Army Corps of Engineers to provide guidance or related training to ensure contracting officers are aware of the regulatory definition of "local area."

USACE RESPONSE: CONCUR with COMMENT.

REQUIRED ACTION – NLT 3 MAR 2019: The USACE Director of Contracting (DOC) will issue a policy alert directing the contracting professionals to follow the guidance at FAR Subpart 26.2 when making the determination on the local area and local firms when contracting for major disaster or emergency assistance activities under the Robert T. Stafford Disaster Relief and Emergency Assistance Act (42 U.S.C. 5150). In addition, the DOC will include a copy of Draft Report GAO-19-281.

RECOMMENDATION 6: The Secretary of the Army should direct the Chief of Engineers under the Assistant Secretary of the Army for Civil Works and the Commanding General of the U.S. Army Corps of Engineers to provide guidance and tools for contracting officials to use to ensure requirements concerning contracting with local vendors, including justification requirements for the use of non-local vendors are consistently met.

USACE RESPONSE: CONCUR with COMMENT.

REQUIRED ACTION – NLT 3 MAR 2019: The USACE Director of Contracting (DOC) will issue a policy alert directing the contracting professionals to follow the guidance at FAR Subpart 26.2 when making the determination on the local area and local firms under the Robert T. Stafford Disaster Relief and Emergency Assistance Act (42 U.S.C. 5150). The alert will encourage the contracting districts and centers to provide guidance and tools for contracting officials to use to ensure requirements concerning contracting with local vendors IAW FAR 26.204, including justification requirements for the use of non-local vendors are consistently met.

RECOMMENDATION 7: The Secretary of the Army should direct the Chief Engineers under the Assistant Secretary of the Army for Civil Works and the Command General of the U.S. Army Corps of Engineers to establish a formal process to solicit input from officials directly involved in the agency's response and recovery following a disaster and to share that input with the Emergency Support Function Leadership Group.

USACE RESPONSE: CONCUR with COMMENT. Action complete.

ACTION – Complete. USACE has a formal process called the USACE Remedial Action Program (CERAP) to solicit input from officials directly involved in the agency's response and recovery following a disaster. These officials include local, state, FEMA, USACE, and other federal agencies. USACE shares specific findings with Emergency Support Function Leadership Group throughout the year and annually during the USACE/FEMA Senior Leaders Seminar. This action satisfies the GAO recommendation intent.

APPENDIX IV: COMMENTS FROM THE DEPARTMENT OF HOMELAND SECURITY

U.S. Department of Homeland Security
Washington, DC 20528

Homeland Security

March 29, 2019

Marie A. Mak
Director, Contracting and National Security Acquisitions
U.S. Government Accountability Office
441 G Street, NW
Washington, DC 20548

Re: Management Response to Draft Report GAO-19-281, "2017 DISASTER CONTRACTING: Actions Needed to Improve the Use of Post-Disaster Contracts to Support Response and Recovery"

Dear Ms. Mak:

Thank you for the opportunity to review and comment on this draft report. The U.S. Department of Homeland Security (DHS) appreciates the U.S. Government Accountability Office's (GAO) work in planning and conducting its review and issuing this report.

Due to unprecedented catastrophic hurricane activity during the 2017 Hurricane Season FEMA experienced a huge increase in Post-Disaster contracts to support response and recovery efforts. Contracting during disasters can pose a unique set of challenges and significant amount of pressure to provide life-sustaining goods and services to survivors as quickly as possible. DHS appreciates GAO's positive recognition of Federal Emergency Management Agency (FEMA) efforts to provide aid to the survivors during this time.

The draft report contained 10 recommendations, including five for DHS and with which the Department concurs with four and nonconcurs with one. Attached find our detailed response to each of these recommendations. Technical comments were provided under separate cover.

Again, thank you for the opportunity to review and comment on this draft report. We look forward to working with you again in the future. Please feel free to contact me if you have any questions.

Sincerely,

JIM H. CRUMPACKER, CIA, CFE
Director
Departmental GAO-OIG Liaison Office

Attachment: Management Response to Recommendations Contained in GAO-19-281

GAO recommended that:

Recommendation 2: Until the NIA code Memorandum of Agreement between the General Services Administration [GSA] and the Departments of Defense [DoD] and Homeland Security is revised, the Secretary of Homeland Security should keep the existing NIA [National Interest Action] codes for disasters (including the code for Hurricane Maria, as well as the 2018 Hurricanes Michael and Florence) open and reopen the NIA codes for Hurricanes Harvey and Irma and direct agencies to retroactively enter NIA codes for contract actions for Hurricanes Harvey and Irma made after June 30 into FPDS-NG [Federal Procurement Data System-Next Generation] to adequately capture contract obligations, to the extent practicable.

Response: Non-concur. DHS and FEMA are bound by the Department's agreement with GSA and DoD unless or until all three signatory agencies agree to revise or suspend the MOA. Extending existing NIA codes or reinstating expired NIA codes without regard to disaster-specific facts applied to the MOA end-dating criteria would have the effect of suspending the MOA's end-dating criteria without the concurrence of all the parties, something DHS will not unilaterally do.

Specifically, as the recommendation relates to the NIA codes for Hurricanes Maria, Michael, Florence, Harvey and Irma:

- FEMA's Office of the Chief Component Procurement Officer (OCCPO) believe that the recommendation to extend the NIA codes for 2018 Hurricanes Michael and Florence goes beyond the scope of this audit, as the facts in the report and the report's scope are limited to the 2017 Hurricanes Harvey, Irma, and Maria (in addition to the California wildfires). The draft report does not contain facts on the NIA codes for 2018 Hurricanes Michael and Florence that support this particular aspect of the recommendation. FEMA is open to considering an agency's request to extend the NIA codes for Hurricanes Michael and Florence if the facts justify such an extension. If such a request is received it will reviewed against the current MOA's end-dating criteria and acted upon, as appropriate.

- FEMA's OCCPO is open to considering an agency's request to further extend the NIA code for 2017 Hurricane Maria, provided the facts justify such an extension. If such a request is received, the hurricane-specific facts will be reviewed against the current MOA's end-dating criteria and the request acted upon, as appropriate.

- Neither DHS nor FEMA can unilaterally direct other agencies to retroactively enter FPDS data for 2017 Hurricanes Harvey and Irma. Even if they could, retroactive FPDS-NG data entry government-wide is not practical and places an unreasonable burden on the same disaster contracting workforce responsible for new time-critical disaster contracts. The draft report does not support the case that any benefits to be gained from retroactive data entry will be outweighed by the burden placed upon contracting officers government-wide.

We request that GAO consider this recommendation resolved and closed as unimplemented.

Recommendation 5: The Commandant of the Coast Guard should provide guidance and tools for contracting officials to use to ensure requirements concerning contracting with local vendors, including justification requirements for the use of non-local vendors, are consistently met.

Response: Concur. The Coast Guard's Office of Procurement Policy and Oversight has already begun to develop the recommended guidance and tools for contracting officers. Estimated Completion Date (ECD): May 31, 2020.

Recommendation 8: The Commandant of the Coast Guard should establish a formal process to solicit input from officials directly involved in the agency's response and recovery following a disaster and to share that input with the Emergency Support Function Leadership Group.

Response: Concur. The Coast Guard's Office of Contingency Preparedness and Exercise Policy is currently reviewing the lessons learned and after action reporting process to inform a 2019 update to the Coast Guard's After Action Program Policy. This office is also leveraging interagency "lessons learned" through the FEMA-led Emergency Support Function Leadership Group to further inform policy improvements. ECD: March 31, 2020.

Recommendation 9: The FEMA Administrator should take the lead to work together with the Coast Guard and the U.S. Army Corps of Engineers to revise the mission assignment policy and related guidance to better incorporate consideration of contracting needs, such as demobilization, and to ensure clear communication of coordination responsibilities related to contracting.

Response: Concur. FEMA's Response Directorate believes the Mission Assignment Policy and Guide as currently written already incorporates the proper language and guidance for project management related to the mission assignment process, including contracting requirements, and meets the intent of this recommendation. This office will work with its Federal partners and others, as appropriate, to better communicate guidance for contracting considerations including for issues such as demobilization. This will include the development of mission assignment project management tools and training for work performed under a mission assignment. To support this effort, the office is currently planning to develop a Mission Assignment Project Manager Guide to help provide the necessary doctrine and guidance to address issues identified by this report.

Action	Estimated Completion Date
Framing and Drafting of Mission Assignment Project Manager Guide	November 30, 2019
Targeted Review of Guide	January 31, 2020
Doctrine and Policy Compliance	March 31, 2020
Final Clearance	May 29, 2020
Socialization	June 30, 2020

Overall ECD: June 30, 2020

Recommendation 10: The FEMA Administrator should assess its workforce needs – including staffing levels, mission needs, and skill gaps – for contracting staff, to include regional offices and DART; and develop a plan, including timelines, to address any gaps.

Response: Concur. FEMA's Office of the Chief Component Procurement Officer (OCCPO) assesses its workforce on an annual basis to determine the appropriate number of contract officials needed to meet its mission. At the beginning of each Fiscal Year, an 1102 staffing model exercise is conducted by all DHS contracting activities. The exercise is based on the current inventory of GS-1102 personnel, number of contract actions, and hours performed (direct and indirect) by GS-1102 personnel. The calculation produced by the model assists management by identifying the number of personnel needed to perform the contract actions. FEMA OCCPO completed its staffing model exercise on January 28, 2019, with a staffing model calculation of 243 Full Time Equivalents (FTEs) needed to meet its mission. The current inventory of the GS-1102 workforce at FEMA's OCCPO is 152 FTEs.

FEMA's OCCPO has utilized the Office of Management and Budget Direct Hiring Authority granted for the GS-1102 series, and recently filled some of its vacancies to include two senior leadership positions. Currently, FEMA's OCCPO has 72 vacant positions, including positions with the Disaster Acquisition Response Team, two key leadership positions at Headquarters, contracting officers and contract specialist positions at various locations, and program analysts.

To address its immediate skill gaps and personnel needs, OCCPO has entered into a contract for acquisition support services for additional personnel to temporarily fill the gaps found during the assessment. In addition, OCCPO plans to hire Cadre of On-Call Response and Recovery Employees to provide dedicated support during disasters.

ECD: September 30, 2019

APPENDIX V: COMMENTS FROM THE GENERAL SERVICES ADMINISTRATION

GSA

The Administrator

April 2, 2019

The Honorable Gene L. Dodaro
Comptroller General of the United States
U.S. Government Accountability Office
Washington, DC 20548

Dear Mr. Dodaro:

The U.S. General Services Administration (GSA) appreciates the opportunity to review and comment on the Government Accountability Office (GAO) draft report titled *2017 Disaster Contracting: Actions Needed to Improve the Use of Post-Disaster Contracts to Support Response and Recovery* (GAO-19-281).

There is one recommendation addressed to the Administrator of General Services:

- The Administrator of General Services, in coordination with the Secretaries of Defense and Homeland Security, should jointly revisit and assess the extent to which the criteria in the 2018 NIA [National Interest Action] code Memorandum of Agreement, including criteria for closing NIA codes, meet long-term visibility needs for high visibility events and account for the needs of users, such as FEMA, other agencies, and the Congress. At a minimum, the agreement should include criteria that takes into account the roles of the federal agencies involved in response and recovery and provide a process that ensures consistent consideration and implementation of the criteria. (Recommendation 1)

GSA concurs with this recommendation and understands the need to regularly review the NIA code management Memorandum of Agreement (MOA) and make revisions as applicable. GSA is committed to reviewing the MOA as soon as practicable with our MOA counterparts at the Departments of Homeland Security and Defense.

Sincerely,

Emily W. Murphy
Administrator

cc: Ms. Marie A. Mak, Director, Contracting and National Security Acquisitions, GAO

APPENDIX VI: ACCESSIBLE DATA

Data Tables

Data Table for Known Post-Disaster Contract Obligations in Response to the 2017 Disasters, in Fiscal Year 2018 Dollars, as of June 30, 2018 dollars in millions

Agency	California wildfires	Hurricane Harvey	Hurricane Irma	Hurricane Maria
U.S. Army Corps of Engineers (USACE)	252.2	70.1	46.1	1867.8
Federal Emergency Management Agency (FEMA)	14.1	358.5	336.3	858.9
Defense Logistics Agency (DLA)		95.4	60	160.3
U.S. Coast Guard		26.3	56.4	63.2
All other agencies		121.2	184.5	342.7

Data Table for Figure 6: Known Federal Contract Obligations in Support of the Three 2017 Hurricanes and California Wildfires, as of June 30, 2018, in Fiscal Year 2018 Dollars (in millions)

Event	Post Disaster Contract Obligations	Advance Contract Obligations
California wildfires	266.3	663.0
Hurricane Harvey	671.5	1160.0
Hurricane Irma	683.4	595.8
Hurricane Maria	3292.8	2122.3

Data Table for Figure 7: Obligations on Post-Disaster Contracts in Support of the 2017 Disasters by Agency, as of June 30, 2018, in Fiscal Year 2018 Dollars (in millions)

Agency	California Wildfires	Hurricane Harvey	Hurricane Irma	Hurrican e Maria
Department of Defense (DOD) - U.S. Army Corps of Engineers	252.2	70.1	46.1	1867.8
Department of Homeland Security (DHS) - Federal Emergency Management Agency	14.1	358.5	336.3	858.9
DOD - Defense Logistics Agency		95.4	60.0	160.3
DHS - U.S. Coast Guard		26.3	56.4	63.2
All other agencies		121.2	184.5	342.7

Data Table for Figure 8: Obligations on Post-Disaster Contracts for Products and Services by 2017 Disaster, as of June 30, 2018, in Fiscal Year 2018 Dollars (in millions)

Event	Services	Products
California wildfires	261.4	4.8
Hurricane Harvey	344.1	327.4
Hurricane Irma	459.7	223.7
Hurricane Maria	2041.2	1251.6

Data Table for Figure 9: Top Post-Disaster Services by 2017 Disaster, as of June 30, 2018, in Fiscal Year 2018 Dollars (in Millions)

Product	California wildfires	Hurricane Harvey	Hurricane Irma	Hurricane Maria	Total Obligations
Maintenance, Alteration/Repair, non-buildings		65.8	110.2	1016.7	1192.7
Management Support Services	0.08	78.2	111.5	199.1	388.9
Housekeeping Services	252.3	16.1	4	82.4	354.8
Construction of Buildings		0.6	5.7	293	299.3
Environmental Systems Protection		17.6	65.6	76.1	159.3
All Other Services	9.1	165.8	162.6	373.9	711.4

Data Table for Figure 10: Top Post-Disaster Products by 2017 Disaster, as of June 30, 2018, in Fiscal Year 2018 Dollars (in Millions)

Product	California wildfires	Hurricane Harvey	Hurricane Irma	Hurricane Maria	Total Obligations
Electric Wire, Power Distribution Equipment	0.02	0.2	9	521.9	531.1
Subsistence (Food and Water)		171.4	83.3	244.8	499.5
Construction and Building Material		1.1	13.8	152.7	167.6
Ships, Small Craft, Pontoon, Docks		0.002	25.1	126.9	151.9
Construction/ Mine/ Excavate/ Highway Equipment			10.1	85.7	95.7
All Other Products	4.8	154.7	82.4	119.7	361.6

Data Table for Figure 11: Contract Obligations over Time for Prior Hurricanes and the 2017 Hurricanes, in Fiscal Year 2018 Dollars (dollars in millions)

Time after hurricane landfall	Hurricane Katrina (2005)	Hurricane Rita (2005)	Hurricane Wilma (2005)	Hurricane Ike (2008)	Hurricane Sandy (2012)	Hurricane Harvey (2017)	Hurricane Irma (2017)	Hurricane Maria (2017)
3 months	8436.5	719.1	88.1	779.1	724.5	1432.5	990.3	3422.4
6 months	3245.6	247	82	107.9	153.5	188.6	161.7	979
9 months	3460.7	120.9	96.4	120.1	245	150.8	99.9	1015.3
1 year	1881.7	188.5	352.4	149.6	485.3	0	0	2
2 years	4547.7	152.6	671.5	100.6	687.9	0	0	0
3 years	1386.3	67.5	374	107.2	462.7	0	0	0
4 years	632	51.1	0.2	5.2	195.8	0	0	0
4+ years	1324.8	162	67	59.3	184	0	0	0

INDEX

D

E

H

I

K

L

M

N

O

P

Q

R

S

T

U

V

W

Related Nova Publications

Emergency Management: An Overview and Issues for Congress

Editor: Stephanie Padbury

Series: Natural Disaster Research, Prediction and Mitigation

Book Description: After a flood, people are often uncertain if their eligibility for federal disaster assistance is linked to any way to whether or not they have flood insurance. The first two chapters in this book provide an overview of the assistance available to individuals and households following a flood and provides links to more comprehensive guidance on both flood insurance and disaster assistance.

Softcover ISBN: 978-1-53614-101-6
Retail Price: $82

Weathering the Storm: A State and Local Perspective on Emergency Management

Author: Anthony Rigg

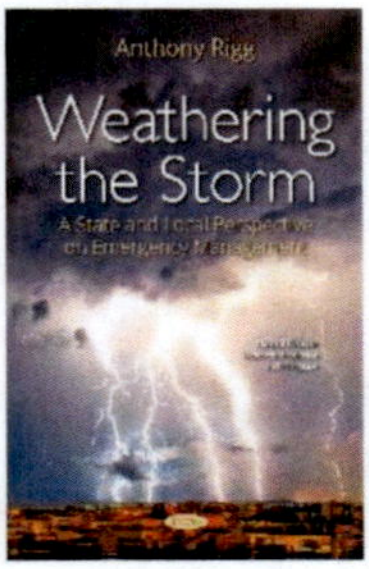

Series: Natural Disaster Research, Prediction and Mitigation

Book Description: This book is the hearing that took place in front of the U.S. House of Representatives on June 10th 2011. It provides a State and Local Perspective on Emergency Management, specifically on the efforts of State, local and non-government organizations to prepare for and respond to natural disaster, terrorist attacks and other emergencies.

Softcover ISBN: 978-1-53613-788-0
Retail Price: $79

To see a complete list of Nova publications, please visit our website at www.novapublishers.com